D0142411

Mastery of Nature

Mastery of Nature

Promises and Prospects

Edited by Svetozar Y. Minkov
and Bernhardt L. Trout

PENN

UNIVERSITY OF PENNSYLVANIA PRESS

PHILADELPHIA

Published by
University of Pennsylvania Press
Philadelphia, Pennsylvania 19104-4112
www.upenn.edu/pennpress

Printed in the United States of America
on acid-free paper

10 9 8 7 6 5 4 3 2 1

A Cataloging-in-Publication record is available from the Library of Congress

ISBN: 978-0-8122-4993-4

One wonders at those [things] which happen according to nature, insofar as one is ignorant of the cause, and at those [things] which happen contrary to nature, insofar as they come into being through art with regard to what is beneficial to human beings. For in many [cases], nature makes the opposite with regard to what is useful to us.

—Aristotle, *The Mechanics*

Nature's bequest gives nothing, but doth lend,
And being frank she lends to those are free.

—Shakespeare, "Sonnet 4"

Contents

Preface

Ralph Lerner

Even imagining such an outcome as mastering nature already bespeaks a certain stance of a human being toward everything around him. That stance might be a result of some reflection or perhaps, rather, little more than an assertion of radical independence: "Let us build us a city, and a tower, with its top in heaven, and let us make us a name" (Gen. 11:4). Conceiving and promoting a *program* to master nature goes even further. That suggests the intervention of some projectors with a pronounced philosophic cast of mind. It is to such individuals that the contributors to this volume direct their readers' attention. In this preface, I consider a different point of departure, one inspired by Wordsworth.

It was long held in the field of human embryology that ontogeny recapitulates phylogeny. Might some analogous "recapitulation" take place in the moral awakening of an individual? Poets, mythmakers, and others have repeatedly tried to reimagine or recover our earliest awareness of the larger world in which we find ourselves. Just as surely as a newborn's physical gestation continues long after it has been delivered from its mother's womb, so too may its psychic awareness be compared to a universe that expands, or at least alters, over time. At first, an infant passing into childhood is enveloped by a world of wonders.

> There was a time when meadow, grove, and stream,
> The earth, and every common sight,
> To me did seem
> Appareled in celestial light,
> The glory and the freshness of a dream.
>
> (Wordsworth, "Ode: Intimations of Immortality," 1–5)

Later, while most of the earth still lay in profound darkness at night, a biped could raise its head and stand in awe of the display in a cloudless sky.

> The heavens declare the glory of God,
> And the firmament showeth his handiwork.
>
> (Ps. 19:1)

Yet this same Psalmist who is struck by the pettiness of man in the context of this cosmic spectacle and is led to wonder that God would even be "mindful" of him (Ps. 8:3–4) is all the more grateful for God's investing man with "dominion."

> Thou hast made him to have dominion over the works of thy hands;
> Thou hast put all things under his feet.
>
> (Ps. 8:6)

This assertion is, perhaps, a fair inference from the passage in Genesis where God brings every beast of the field and every fowl of the air for Adam's inspection—"to see what he would call them; and whatsoever the man would call every living creature, that was the name thereof" (Gen. 2:19). Does the very act of naming imply mastery? Yet Adam and Eve were and remained herbivores. It was not until God's covenant with Noah after the Flood that man was authorized to view every living thing on the earth, in the air, and in the seas as fair game, as a potential meal. Now it could rightly be said that "the fear of you and the dread of you shall be upon . . . every moving thing that liveth" (Gen. 9:2–3). Here is mastery indeed; but in promising plenitude and exhibiting divine beneficence, there is little to suggest that man's relation to the physical, natural world is determinedly adversarial. Milton's primal pair, bearing new curses, might have been somewhat consoled anticipating some such future upon being expelled from Eden.

> Some natural tears they dropped, but wiped them soon:
> The world was all before them, where to choose
> Their place of rest, and Providence their guide;
> They, hand in hand, with wandering steps and slow
> Through Eden took their solitary way.
>
> (*Paradise Lost*, XII. 645–649)

We are pleased to believe that we will be well provided for in our necessities by Mother Nature.

This commonplace of children's books of yesteryear is both quaint and unsustainable in the face of further life experience. The frequency of plagues, floods, droughts, and other natural disasters suggests powerfully one of two thoughts: *either* that such sudden vicissitudes had to do with *our* shortcomings and backsliding—a doctrine propounded in Leviticus and Deuteronomy, and much invoked by Congregationalist ministers in Massachusetts Bay Colony even beyond the seventeenth century—*or* that man cannot rely, *dare* not rely on the presumed goodness of a caring mother. Man has to act for himself and reshape nature to serve his needs—and, later, his wants.

Here is the proclamation of a new declaration of independence from earlier beliefs. A new stance toward the natural world as a whole is being proposed, one that promises relief from pain and want and misery. Not only had a far-seeing elite to be persuaded that this was not pie in the sky, but an attainable goal well within reach, if only we set our minds to the task. Equally important was the broad campaign to persuade mankind at large that this was a project worthy of their support and eager anticipation. In raising popular expectations of greater longevity, comfort, and health, the seventeenth-century projectors of this concerted effort to bring nature to heel succeeded in replacing one poetic vision with another. The magnitude of their actual success in matters of longevity, comfort, and health has effectively left that earlier vision by now both largely unnoticed and unmissed.

> The world is too much with us; late and soon,
> Getting and spending, we lay waste our powers;
> Little we see in Nature that is ours;
> We have given our hearts away, a sordid boon!
>
> (Wordsworth, "The World Is Too Much with Us," 1–4)

Introduction

Daniel A. Doneson, Svetozar Y. Minkov, and Bernhardt L. Trout

It would be a bold soul, or at any rate a contented one, who holds that the mastery of nature is unproblematic; or that a world in which the idea of the mastery of nature reigns is simply a happy one. Nevertheless, the boldness or contentedness inherent in the belief that mastery of nature should be pursued with all effort, and that we should continuously push the world toward a future with greater and greater mastery of nature, might be deflated when we are reminded of the disastrous possibilities to which such thoughts might lead. More and more ways for the annihilation of the human race are enabled by ever more powerful military technologies, weapons, robots, or genetic modifications, for example, and from their civilian counterparts. Environmental cataclysms of our own making could render our planet uninhabitable. Technology could be used for the inexorable control of our actions, or technocracy could lead to the enslavement of the human race. The indiscriminate use of technology could lead to the alteration of our biology, such that we become what today we would not even recognize. Or perhaps, we would just become more and more disoriented in an ever increasingly technological world, such that we do not even know where to turn for happiness. Such thoughts should give even the boldest and most contented pause and even lead them to pose the question: what are the alternatives?

Perhaps, however, we should take a step back and investigate what mastery of nature is. For mastery of nature is not merely technology or its products. If it were, then mastery of nature would be for the most part unproblematic. It would be part and parcel of human existence, something necessary to provide us with what we need to live: food, shelter, and clothing, together with the various tools needed to secure them. Certainly, people would want more than just the bare necessities, so to the extent that they could, they would procure wine and spices, palaces and fine garments, jewelry, sculptures, and other objects of beauty as they could obtain them. They would not, however, spend the better part of their existence aiming to master nature. But mastery of nature is not just one activity out of many, such as pottery or sports or politics. It is all-encompassing, directing us in ways that are difficult, if not impossible to resist. The potter must now use the latest materials and machines, the athlete must use the latest equipment, and the politician must promote mastery of nature. In addition to being omnipresent, mastery of nature involves a universally applicable approach to all things under the sun and beyond the sun.

As such, mastery of nature is also a moral and a political horizon. Perhaps, as Heidegger has suggested, the world is transformed into "standing reserve," and therein, all human effort is turned toward exploiting the standing reserve. Countries spend vast resources to do so, and individuals are compelled to do so more and more in just about every aspect of their lives. Whether seeing ourselves as inconsequential specks in an unfathomably vast (but quantifiable) universe, or focusing with greater and greater effort on less and less consequential activities, mastery of nature all but consumes our mind and our actions, and not necessarily in the best ways.

Thinking of mastery of nature in this way, we may wish to ask about its origins. Such questions would not be merely idle ones, for while we tend to forget about foundations and focus on the magnificent structures themselves, the structures are held up by the foundations, and the foundations give insight into the structures that may not be evident from studying only the structures. Mastery of nature as such seems to have begun in the West, around the time of the Renaissance, as a response to a crisis in religion and thought. While there are many important thinkers who addressed the crisis, this volume focuses on the handful of thinkers who most directly and comprehensively laid out the aims, as well as the attendant moral and political dispositions and forms, that are part and parcel of mastery of nature: Machiavelli, Bacon, Hobbes, and Descartes.

These thinkers directed us toward conquering fortune by harnessing nature to do our bidding. In making their case, they directly opposed their new thought to that of the ancients, which they characterized as "fruitful of controversies, but barren of works." Exploring their thought, we can see that mastery of nature is inherently a philosophical and political project, based on a new notion of virtue in which human beings would be redirected from theoretical speculation toward productive works. Mastery of nature promised to conquer disease and other hardships and create plenty, thus promoting peace on earth. Princes, kings, and princesses were solicited to support the project, and the people were charmed with the rewards that they would obtain by it. The project took off and, within a matter of centuries, was so successful in fulfilling wants that all the world's politicians came to promote it. In time, a new understanding of the first principles was developed, in addition to a new cosmology and cosmogony. We cannot now be unaffected by its products: biotechnology, nanotechnology, chemical technology, information technology, and related disciplines influence practically every aspect of our lives. Systematic procedures, which themselves are the application of tools of mastery of nature to politics, dominate a vast part of our lives, and ever expanding government uses these tools to control just about our every activity. Thus, understanding mastery of nature is essential in addressing the question of human freedom. Does the charity or generosity of mastery of nature, with everything it gives to us to increase our freedom, ultimately lead to our enslavement?

Beyond the problem of mastery of nature and human freedom is the problem of mastery of nature and morality. Does mastery of nature presume a certain moral disposition? Can the methods employed to master nature say anything themselves about morality? If the answer to the latter question is "no," then any presumed moral disposition is problematic, for mastery of nature could not continue to justify itself via the thought of its own making. Forgetfulness of the original moral disposition would ensue, as human beings adopt whatever notion of morality is consistent with mastery of nature. Even environmentalists, who might seem to be the last bastion of anti-mastery of nature, promote the use of mastery of nature to solve the problems created by mastery of nature.

Furthermore, what of the assumed connection between wisdom and power in mastery of nature? If anything, mastery of nature directs human thought away from questioning that assumption and toward increasing power over nature, and while increasing our power, it does not increase our understanding of how best to use that power. It directs us away from gaining "knowledge of ignorance," for there seems to be a trade-off: use our time to pursue power or to pursue wisdom.

Many thinkers tried to address these problems in various ways, including Rousseau, Kant, Hegel, Nietzsche, and Heidegger, and despite significant influence of these thinkers, mastery of nature continues forward with all force. Mastery of nature reigns supreme with all of its promises and problems, and there is no serious dissent from it. Moreover, there is, perhaps because of its successes (and because of the failures of subsequent thought to unhinge it), very little investigation into mastery of nature. The problems, however, remain whether they are addressed seriously or not. Because of this, it would not be unprofitable to start to think them through. This volume, thus, aims to invigorate thought about mastery of nature. Given the vastness of the subject, it can be only a beginning, the starting point for the elucidation of the problems that it entails and its prospects.

* * *

Has the mastery of nature expanded our vision or constricted the horizon of science to the merely manipulable or useful? Has modern science made genuine theoretical discoveries or has it abandoned the search for pure theory? Are the mastery of nature and modern science instances of an irresistible reductionist fate that dissolves the continuity, unity, or wholeness of natural experiences and beings? Or will a fully developed modern science vindicate common sense while enriching it, healing the sciences in their self-inflicted overspecialization and infusing confidence in the now despaired-of rational reflection on the human good?

Whatever the judgment about its ultimate value, the theoretical and practical issues of the mastery of nature point to an elementary philosophical question: is there a fixed human nature? And if there is a fixed human nature, is that something to accept and defer to or is it something that could be radically modified or even abandoned? Abandoning human nature as a standard may seem Sisyphean or absurd because the standard for our objections to human nature would have to be the promptings of human nature itself. Yet how would one argue that human nature, as we know it, is better than any new species one might produce technologically? Perhaps, then, it is possible to rebel against human nature on the basis of a reasoned-out goal (such as justice) or a painful feeling (a tension within human life). But even those ostensibly "Archimedean-point" goals would be implicitly rooted in a notion of human nature, even as that nature is being remade or reprogrammed.

Such questions and reflections arise from, but also bring in their train, doubts about the status of the mastery project. In response to these doubts, we see various attempts to arrest the disorienting or self-undermining aspects of the project. What we seem to encounter, however, are *unsuccessful* efforts to restore something of the dignity of contemplation (e.g., teleology in science) and of moral virtue. These attempts are unsuccessful in part because they do not reassert the primacy of contemplation with the rigor of Spinoza, whose own doctrine in favor of liberal democracy also helps along the mastery-oriented march of history. As becomes evident in Rousseau, the critique of the mastery project thus becomes the vehicle for its radicalization and estrangement from the fundamental questions of human life. We say "radicalization" since, while the master-founder Bacon might have said, "let people chase inadequate images of happiness through money and success," Rousseau's very warnings against the Enlightenment use Enlightenment means and, hence, willy-nilly appeal to everyone, adding to the obfuscation of the distinction between philosophers and non-philosophers. While he himself understood profoundly the difference between the social good and the individual good, Rousseau ends up inflaming the hope for happiness in everyone's heart by rejecting technology and capitalism in favor of serene resignation (see Goethe's letter to Zelter, June 6, 1825: "Nowadays everything is ultra . . . no one knows himself . . . Wealth and rapidity are what the world admires, and what everyone strives to attain. Railways, quick mails, steamships, and every possible kind of facility in the way of communication are what the educated world has in view, that it may over-educate itself, and thereby continue in a state of mediocrity").

While one may think that the politicization of reason goes with mastery and with a rebellion against, or a lament of the nonexistence of, natural ends, someone like Kant may be said to attempt to salvage teleological satisfaction *precisely* in understanding reason as practical at its core. Kant attempts this solution by attributing metaphysical significance to the moral project of the "relief of man's estate"—a moral significance already rhetorically built into Bacon's advertisement of the project. Yet Kant's turn proves only apparently pre-modern but is in fact another expression of assertive or groundless modern self-grounding. In the end, however, Kant finds a basis for objecting to human mastery in a traditional, medieval source: his moral meliorism ends up in religion, the assertion of man's "radical evil," a gratitude for the "fortunate elusiveness" of knowledge and "the thrill of not knowing" if one is actor or patient. Hegel also looks for an anchor for philosophy of nature in the problem of consciousness (individual and communal) and has hopes for the completion of the development of consciousness in absolute science. But while Hegel may reasonably criticize some reductionist aspects of modern science and attempt to restore a connection to phenomena, his philosophy of nature remains inadequate. His neo-Platonic phenomenology or metaphysics, based as it is on an implicit "Cartesian" starting point, cannot correct for scientific reductionism.

From a Nietzschean-Heideggerian perspective, one might argue that precisely if modern science proves technically successful, its side effects—decadence, mechanistic coldness, and homogenization—would have to be neutralized. A global society may require a new world culture, and such a cohesive culture may not be sustainable on the basis of any of the traditional religions. Perhaps a syncretistic fusion of religions would come to the fore, but such eclecticism might be too weak-kneed and undemanding to elicit the potential of, and bring together, the new global humanity. We might look, then, to the emergence of new theophany to fulfill the function of authoritative guidance. In that case, the hopes for merely revising or supplementing the modern project would yield to the emergent *mythoi* of a wholly new beginning, or at least a radical recovery of something long lost. We should also be open to the appearance of new philosophers (even in unlikeliest of places and the unlikeliest of times) who might conceive of the human situation in dramatically new ways. And, perhaps instructed by these new guideposts, which may not even require new philosophers but only powerful restorers of old philosophies, the mastery project would be reinvigorated in its theoretical or scientific dimensions, or alternatively dispensed with altogether.

The capacity of Enlightenment thought for self-criticism leaves it ambiguous as to whether the challenges and confusions of late modernity are due to defects inherent in the Enlightenment or to the misapplication (or insufficient application) of principles articulated at the outset of the mastery project. In the historical unfolding of the Enlightenment, the idea of the mastery of nature alternates between receiving large injections of hope and suffering withering setbacks. Despite repeated criticism, and despite convincing declarations of crises, the idea, or the hope in the idea, of the mastery of nature shows no signs of disappearing from the human consciousness. It may still be the most powerful global force. If it loses steam, it seems that this would be due not to an aggressive alternative ready to take its place, but because of the complacency of its own success. If the prospects of the project are uncertain, that uncertainty is compounded, if also concealed, mostly by the fact that the project is going along virtually uncontested. The world of radical Islam may be its only serious external challenge, but even its leaders and followers are not averse to learning about and using the latest technology, even if they do not seem to originate any.

It may be unreasonable to fault the early modern founders of the mastery project for not anticipating a late modern loss of direction due precisely to the excessive success of their enterprise. The founders would be even more blameless—they would not even be "founders"—if they were the instruments of providence, history, or other impersonal forces. But we must also not overlook a third possibility, in which the mastery project declines because of defects or limits which the project's founders fully anticipated, if not preplanned. In that case, if the project goes astray or runs its course, such exinanition would be due not to an insufficient application of its principles, nor to a miscalculation by its founders, but to a kind of built-in, expected, and even counted-on obsolescence. There are indications in the works of Machiavelli and Bacon that they anticipated the future emergence of new sects or new dominant "ideologies"—necessarily indeterminate in advance—as a reflection of the evolving need for ways to manage the difference between mind or philosophy and the realm of opinion.

The gulf between ancient and modern conceptions of that difference appears further bridged upon discovering that the ancient thinker Xenophon may have regarded the question of support or resistance to a mastery of nature project as a prudential matter. This might explain why Bacon's or Franklin's realistic conception of morality and politics, while similar to Xenophon's, has led to such different conclusions regarding the political desirability of a mastery project. Bridging the gap between such apparently different thinkers as Xenophon and Bacon from the other direction, one may find in Bacon an argument that the moral sense of deserving and justice to be a permanent feature of the human psyche. Hence, there will always be moral confusions (as well as a longing for piety), and late modern diagnosticians such as Nietzsche or Heidegger ought not to be overly anxious that the mastery project will unmoor itself into freewheeling will to power, "will to will," or nihilism.

We suspect that, were they alive today, most of the philosophers discussed in the chapters below would have written differently from the way they did during their own time. This is not to say that their works are historically contingent, or that they were merely tracts for their times. On the contrary, as the essays in this volume show, despite any contingencies in their rhetoric, these philosophers give us fundamental insight into the relationship between mastery of nature and the human condition as such. They continually reinforce the never-abated necessity that we have to think for ourselves, situated where *we* are, even when, or precisely when, we are attempting to learn from those much wiser than we are. In trying to achieve a degree of thoughtfulness for our own lives, we can perhaps contribute a measure of thoughtfulness to the broader world. Thus, despite or because of the power and lure of its accomplishments, the severity of its dangers, or the character of its self-doubt, the mastery project has not disposed of what is most pressing and most important for each of us—pursuing an answer to the question, "How should I live?"

* * *

Harvey Mansfield offers a reading of chapter 15 of Machiavelli's *Prince* and its call to go to the "effectual truth" (*verità effettuale*) as opposed to the imagined truth stated in professions of goodness. In this way, Mansfield uncovers the establishment of "fact," which frees science to imagine the project of mastery, "the use of theory to improve human art, science, and technology, with the end to improve human life." Mansfield's Machiavelli learns from Christianity, which also began from human suffering. He even appropriates it as he seeks to overcome it. The "fact" that lies at the root of modern science would replace the religious attempts, born of the "imagination," to seek mastery over nature and chance. While Machiavelli may go a long way toward scientific determinism, he does not go all the way. By retaining the need for good fortune, he holds to human freedom and virtue in the management of fortune.

Minkov's Bacon then brings Machiavelli's epoch-making intervention in the history of philosophy to the forefront. But the relation between these two thinkers is not simple. Bacon all but stated the modern purpose: a universal state, though not a universal state of free and equal citizens. Moreover, he is the first one to make a case for the *scientific* project for mastery of nature in its various dimensions. In uncovering Bacon's intention, Minkov turns to his long neglected *On Wisdom of the Ancients* as Bacon's most sustained treatment of the mastery of nature. Minkov's Bacon deftly employs pity, self-abasement, and devotion to teach the new humanitarian ends. In this way, Minkov's Bacon finds Christianity even more useful, and more easily exploited for non-Christian ends, than Mansfield's Machiavelli. Thus, Bacon suggests that science does Christianity one better: scientific, technological philosophy is the most pious or charitable activity.

In "Hobbes on Nature and Its Conquest," Devin Stauffer shows that Hobbes's thoughts regarding the mastery of nature are more complicated than they at first seem. It is true that Hobbes regarded the human construction of the mighty Leviathan—the commonwealth—as the rational response to man's dire natural condition; but it is also necessary to take into account the more skeptical aspects of Hobbes's natural philosophy in order to appreciate the blend of confidence and doubt that drove him to join in the early modern call for the conquest of nature.

Stuart Warner's essay explores elements of Descartes's notions of mastery and nature through a textual analysis of some hitherto unexplored themes and dimensions of his first major work, the *Discourse on Method.* In particular, it reveals the dual structure of the *Discourse*, insofar as the first half of the work is shaped by a concern to explore common opinion and convention, and the second half turns away from that and toward nature. Additionally, the essay places the ubiquitous irony of the work in sharp relief.

Montesquieu famously made the case for commerce, but Diana Schaub points out that he made the nearly identical case for science: they both cure "destructive prejudices." However, Schaub's Montesquieu is not simply content with the benefits of modern science; he was equally concerned with its moral and political ramifications. How compatible is modern science with liberty? Schaub's Montesquieu laments the alienating effects of Spinoza's reductionist materialism, since he fears it undermines the distinctions necessary to sustain political life. In his great book, it is music and commerce that sweeten or soften (*adoucir*) human action, and in her arresting phrase, "commerce is the music of modernity." Schaub suggests for Montesquieu that "commerce is the shape in which science (as practical technology) ordinarily manifests itself."

Jerry Weinberger uncovers the connections between the thought of Francis Bacon and that of Benjamin Franklin, in particular on the intersections of mastery of nature and religion. In doing so, he shows how Bacon and Franklin conceive of morality and how their confidence in the ineradicability of less-than-rational moral beliefs emboldened them in their projects for technological improvement.

Switching to the ancients, Robert Bartlett contrasts the modern philosopher-scientists as seeking out "the mastery and possession of nature" or the "conquest" of nature, whereas the ancient philosopher-scientists were evidently content just to look on, observe, or "contemplate" nature. To begin to understand this primacy of "contemplative" (or theoretical) virtue for classical thought, Robert Bartlett's essay considers the moral and political thought of Plato and Aristotle. In particular, it examines the case they make for contemplative virtue, especially in its relation to our concern for happiness. This classical analysis of virtue and happiness is alive to the kinds of criticisms that would come to be leveled against it by the modern thinkers—by Thomas Hobbes, for example—and yet Aristotle, like Plato, insisted on virtue's centrality. Why? What price did they think we would pay, were we to abandon, or attempt to abandon, that concern for a more "realistic" one? The answer to this question is in turn bound up with the status of the longing for eternity in the human soul, a longing the modern thinkers tended to think was either inessential to us or—increasingly in our own time—satisfiable. But this, according to Bartlett, Plato and Aristotle denied.

Christopher Nadon presents Xenophon's writings as showing a keen awareness that political life becomes possible for human beings only through the conquest or mastery of nature. Yet the character of that political life also requires that nature, or something, act as a limit standing above or beyond it. While the conquest of nature should not then become the explicit object of a political program or project, circumstances may in certain cases justify a departure from this rule.

Paul Ludwig argues that while Lucretius shares with modern science an atomic view of nature, he would have disagreed with the modern project on the important question of the medical and technological attempt to defeat death. By subordinating all other questions to the overarching inquiry into how to live a truly pleasant life, Lucretius raises problems about what is meant by laws of nature, whether wholes are as real as their component matter, and whether form or matter better describes a thing.

In Arthur Melzer's view, Rousseau opposes the modern project for the conquest of nature as something harmful to both the well-ordered society and the happy individual—harmful because destructive of moderation, that great classical ideal. Rousseau's well-ordered society, the Spartan martial republic, requires citizens who have strictly moderated their selfish, acquisitive desires in order to make possible an ardent, self-sacrificing patriotism. But private individuals seeking a personal happiness also require moderation in the still deeper sense of knowing and accepting their human limits, learning how to die, resigning themselves to necessity. The modern technological project destroys this essential moderation because, on the material level, it stimulates acquisitiveness and a coddling pursuit of comfort, and on the larger cultural level, it promotes a spirit of rebellion and conquest and spurns the posture of resignation.

For Richard Velkley, Kant's philosophy offers an equivocal endorsement of mastery of nature. In philosophy's humanitarian project, human self-mastery has precedence over the use of nature for human benefit. The human is called "lord of nature" in a system of purposes, but only so far as the human will pursues a final end beyond nature. Moreover, teleological views of reason and nature are in permanent tension with the mathematical account of nature.

Michael Gillespie argues that Hegel's natural science is not as weak as is ordinarily believed, that it rightly resists the rabid reductionism of modern science since Descartes, and that it attempts to provide a phenomenological or experiential basis to science that we today often find lacking. Yet Gillespie finds that Hegel's account is still too conditioned by his acceptance of the Kantian notion of consciousness as central and, for that reason, has been superseded by those willing to investigate the foundations of consciousness itself.

Lise van Boxel's Nietzsche thinks that the human being, and in fact every being, is a kind of becoming. To see a being as Nietzsche sees it is therefore to see its genealogy. His account of beings, and of the human being in particular, departs radically from the theological and scientific or philosophic traditions. Nietzsche argues that these traditions are fundamentally the same in that both are determined by the conflation of two prejudices—namely, the moral prejudice and the theological prejudice. The compound prejudice yields the belief that there is an eternal, unchanging, and pure being or good that rules and defines the cosmos and all beings, including the human being. Whereas Nietzsche's account accords with the theory of evolution or generation, the moral-theological prejudice cannot accommodate the becoming that is inherent in what we are.

Mark Blitz describes Martin Heidegger's view of technology, for in his judgment, technological understanding is the source of the possibility of the mastery of nature. He examines the roots of Heidegger's view of technology in his interpretation of being and of human being generally. He then discusses ways that Heidegger attempts to deal with the problem of technology and sketches alternatives to his understanding that do not ignore his insights. Along the way, Blitz briefly discusses the meaning of "nature," limits to what technology can master, and Heidegger's politics.

Adam Schulman considers the question of whether modern science, despite its turn away from merely speculative inquiry toward the practical goal of conquest or mastery of nature, nevertheless manages to shed a genuine theoretical light on the nature of things. After considering the apparent disagreement between Plato and Aristotle on the availability of a science of nature, Schulman examines two great discoveries of modern theoretical physics—space-time and the quantum of action—in order to show that modern science has indeed given us insights into nature that, by a reasonable standard, amount to causes, principles, and elements in the Aristotelian sense of those terms.

Bernhardt Trout discusses insights that quantum mechanics can bring to political philosophy. His approach is dialectic—revealing, on the one hand, the new challenges that quantum mechanics brings to bear on political philosophy, and, on the other hand, the new openness of quantum mechanics to wholes. In other words, quantum mechanics presents a mechanistic view of nature that is not reductionist and, therefore, is open to viewing political things as having their own intrinsic importance. Thus, there could be a direct connection between the approach to nature in the mastery project as it has evolved and the approach of political philosophy.

PART I

The Project for Mastery

Chapter 1

Machiavelli and the Discovery of Fact

Harvey C. Mansfield

In this chapter I shall address the mastery of nature from the standpoint of the discovery of fact. Strange to say, "fact" had to be discovered—one could even say invented. Fact has a meaning and comes from a concept. It is not as forthright and simple as it pretends to be. It was a momentous step when truth came to be understood as fact, and that step, I propose, was taken by Machiavelli, who was a founder not only in politics but in philosophy or science. It was he who created the world of fact we now inhabit. For some reason, he did not take credit for his creation but decided to remain more or less unseen behind it as the animating guide to its meaning. For the meaning of fact is that it brings us up short; it resists wishful meaning and thus seems to reprove our desire to control the world. Somehow the facts we cannot control are critical to the new intention of modern philosophy and science to control the world and master nature.

How is it possible, then, to conceive the mastery of nature if nature consists of facts? Facts are indomitable, unwilling to be mastered; they are resistance itself. Resistance personified is "stubborn," and so are facts. "Facts are stubborn things," said John Adams in 1770 when defending British soldiers on trial for the Boston Massacre, echoed by Bernard Mandeville's Horatio (1732) to concede the valor of Cromwell's army, both men speaking against their inclination.

The matter of fact is pure matter, without potentiality. One cannot ascend from a fact to its meaning, to what makes it intelligible. What makes a fact perceptible is not a sign of being, of the thing in itself, to which one can reason with the aid of imagination. "You call that a meal?" expresses dissatisfaction with the actual as opposed to the perfect meal, but the actual remains a fact. A fact is not an imperfect version of something perfect. It is what it is, its reality equaling its appearance. So one must accept facts as given, in contrast to works of the imagination.

Instead of perfecting facts, one must isolate them. One must isolate the unintelligible given—and here comes mastery—in order to liberate the intelligence to conceive. If one had to imagine things on the basis of perceptions, by perfecting them, distinguishing carefully what is essential to the thing and what is accidental to it, one would be limited, hampered, by having to take seriously the ordinary appearances of things in order to discover what they "mean," or what they imply. In that view, it would be hard to know which is true, the ordinary meaning that holds for most cases (for example, lead falling faster than feathers) or the extraordinary meaning of the scientist (the two falling at the same rate in a vacuum produced in a laboratory). There would be a loose meaning of every perception for ordinary life and a strict meaning for the adepts. The adepts would spend their time telling ordinary folk that they are deluded—or perhaps worse, leaving them in delusion and not telling them—without being able to guide them to an understanding that offers greater mastery over their lives. There would seem to be one "nature" viewed differently by non-philosophers and philosophers, both trying to make it intelligible as helpful to them but disagreeing as to which view is correct. Nature would always be mixed up with meaning, that is, with convention, nurture, and culture in such a way that nature limits, hampers, or prevents mastery over nature. All mastery would seem to be guided by the nature men want to master. Even in improving nature, men would be subordinate to it.

So, in its drive toward mastery, modern science tries to establish facts. Facts are *data*, given things, which cannot be fiddled or monkeyed with. Nature consists of what is purely given, "almost worthless material," with no promises of anything better, no inclinations to perfection attached to one's perceptions, no "innate ideas" to serve as a guide.[1] The purpose of modern science is to see nature as pure fact as opposed to human wish, and in this it reverts from the Socratic turn to human opinion as the beginning of wisdom: to the original pre-Socratic distinction between nature and convention that sees the two as opposed, dichotomous, each untainted with the other. In the philosophy of Francis Bacon, however, wherein science is being led to modernity, science has to deal with the deceitfulness of nature that makes ordinary folk and philosophers, who take them seriously, want to follow, instead of master, nature. Science must undertake a critique of ordinary perceptions together with a critique (or more likely, a flat rejection without a critique) of Aristotle. Nature is not just sitting there innocently, ready to be mastered; nature must be interrogated, treated with what our time calls "enhanced interrogation techniques," or, to speak plainly, even "tortured."

Nature must be reduced to facts, and not facts merely of ordinary perception, but scientific facts, experimental facts. These prove to be facts you never see, like a feather and a lump of lead falling at the same rate. These are facts invisible to the naked eye; the natural eye has to be aided by artificial aids such as the microscope and telescope. Science exists by looking at things you can't ordinarily see. The visible is less real, less a "fact," than the invisible; the real facts are of unbelievably large or unbelievably small size. To get past convention to find fact, one has to get past conventional vision—though isn't that natural?—to arrive at the full notion of fact. For modern science, especially social science, nature is all or nothing. It is either entirely intelligible as a fact universally, which means without human intervention, or it is not at all intelligible. Is maternal instinct a fact of women's psychology? It either has to be so universally and without variation, and spontaneously without conscious thought, that is, a determinate fact of life without fail, or it is not fact, because it has been corrupted by human interpretation and choice. There is no "inclination," no final cause, of being a mother that could be completed in various ways in different cultures, not universally and exactly but probably and "for the most part" (Aristotle's phrase[2]).

The establishment—it goes beyond mere recognition—of fact frees science from the inclinations and promises of previously so-called nature. It can develop its own imagination, now guided by mastery, the use of theory, to improve human art, science, and technology, with the end being to improve human life. Human life consists of suffering—Christianity was right about that—but it is suffering that can be relieved. Thus modern philosophy, the tutor of science in its youth, can be both empirical, to establish facts, and rational, to conceive theories or models in abstraction from facts. The models can make use of mathematics, a new mathematics newly abstracted from facts, counting abstracted from things counted. Empiricism and rationalism, however apparently opposed, cooperate with one another: empiricism to liberate science from intelligible nature (with its skepticism over the "natures" of things); rationalism to occupy and exploit the field left empty as an intended result.

Machiavelli begins the shift to fact that occurs with modern science. He shows us the original motive for making the change from perception as appearance to perception as fact. This was to oppose, and then appropriate, the enemy to human freedom in religious vision. The trouble with Socratic philosophy, with its intelligible nature resting on perfected perceptions, was that it could be taken over by religious vision. For modern science was not the only force seeking mastery over nature; this was also characteristic of religion, both pagan and Christian but especially the latter. When confronted with Plato's idea of the good, ordinary people can be made to suppose that their good is what will make them secure in the future, immediate and far-off. They want to know what is going to happen to them. So they personify persons, gods or God, who secure "the good" for them, not merely as a thought in the minds of philosophers. This religion is an attempt to master nature, and Machiavelli shows that humans, using prudence and even "science" (*D* 3.39.2[3]), can appropriate Christianity and use its methods against itself.

Machiavelli seems to be interested in politics and not in the science of nature. But it would be much better to say that he approaches the science of nature through politics. His approach can be seen at its most explicit in the famous paragraph that begins chapter 15 of *The Prince*. From this start I will try to show, *in brevissimo tempo*, how he sets the stage for the discovery of fact. That discovery, "fact" by name in its modern sense, seems to have been made by Thomas Hobbes (though the word is used by Francis Bacon), working from Machiavelli but not mentioning him, in establishing the contrast between "fact" and "imagination" presented in his *Leviathan* and forecast by Machiavelli.

We can begin with Machiavelli, the professor of necessity (as opposed to the professors of imaginary good from whom he departs). His appeal to necessity is designed overall to simplify not just our politics and morality but our thinking in general. Necessity will give us access to the truth without having to sort out dialectical disputes or to consult high-minded rationalizations. Yet in "fact"—the word not quite invented but prepared by Machiavelli—necessity is not as simple as it first appears. The last sentence of that paragraph, Machiavelli's clarion call to modern morality and modern politics, is as follows: "Hence it is necessary for a prince, if he wants to maintain himself, to learn to be able not to be good, and to use this and not use it according to necessity." He identifies his departure from "the orders of others" as moving to a new standard of necessity, and he makes it emphatic by using "necessary" twice and in two different meanings. The first is what one is compelled to do; the second is the standard for choice, "according to" which one must act when one appears to have a choice. When not compelled by necessity, it appears, one must choose it (*D* 1.1.4–6). This double meaning is the first item of complexity in necessity: that necessity is not always compelling and does not in every case do away with choice.

Machiavelli gives a reason for adopting the focus of necessity in the exercise of one's choices: "A man who wants to make a profession of good in all regards must come to ruin among so many who are not good" (*P* 15.61). This person could be a political scientist or philosopher like himself, and he immediately mentions the "many who have imagined republics and principalities that have never been seen or known to exist in truth." "A profession of good," the standard Machiavelli departs from, represents a choice made regardless of necessity, even in defiance of necessity, as when one acts, and defends one's action, by professing that it is good, regardless of the sacrifice of one's own well-being and the risk of coming to ruin. Making a sacrifice, taking a risk, is what is known as nobility, though Machiavelli does not mention it here. Machiavelli, to put it mildly, is no friend of the "gentlemen" (*kaloi k'agathoi*, "the noble and good") who are addressed by Plato and Aristotle. Also included in the category of those nobly resisting necessity might be Christian martyrs. Though it may well be true that noble examples are rare, they are impressive and are able to set the standard by which the gentlemen, and the ordinary people who admire them, judge others and themselves. Despite its focus on the noble few, this standard has the ability to make itself universal, encompassing all humans, by taking advantage of human admiration for the best. Machiavelli departs from this standard and creates a new one to replace it.

Now in the old standard, what is the reason for making a *profession* of good, rather than merely *doing* good? Machiavelli implies that the profession is needed and is crucial rather than ornamental. Goodness does not stand on its own unaided; it needs the support of a profession that makes goodness possible or reasonable to attempt. If you are good, will not the wicked gleefully proceed to take advantage of you? You must therefore presuppose a good society, one not in the hands of rascals and rogues, that will make it possible for you to be good without coming to ruin. And the good society must be compatible with human nature, which too must good, and then the goodness of human nature must be compatible with, or comforted by, the goodness of non-human nature, the whole. For what can human goodness accomplish on its own, so to speak, without nature's cooperation? Nature must contribute an environment in which good men can thrive, powerful inclinations toward good in the human soul, and a regularity of motions and seasons permitting good men to live in confidence and understanding rather than fear for survival in blind ignorance.

So Machiavelli rightly extends the required reason behind doing good to a "profession," that is, an explanation of the contextual support, and that profession of good must be "in all regards." The reassurance that morality needs is a profession of the whole, clearly a philosophical task. If Machiavelli is going to dispute the profession of good that philosophers, especially the classic ones Plato and Aristotle, have provided, he will have to cover the same ground in order to show that he is right and they are wrong. He will have to make a profession of necessity in all regards counter to the profession of good in all regards.

Whereas Aristotle starts his *Nicomachean Ethics* from the existence and practice of moral people, implying that morality exists, is viable, and is a going concern that one merely has to examine rather than create,[4] Machiavelli begins this critical passage with a critique of morality, denying that it is viable and asserting that it will bring you to ruin. To ruin! He disagrees with Aristotle that morality exists and adopts the Christian view of the sinfulness of the world, but he seems to foreclose the redemption in the next world promised by Christianity. The redeemer he promises in *The Prince* is a worldly one for Italy (*P* 26.105). In *The Prince* and the *Discourses on Livy*, Machiavelli speaks of "the world" rather than "this world," which implies another world beyond this one.

Necessity, then, has the character of a presumption, rivaling the presumption of the good. As a presumption, necessity is not a determination that, in each case, one who chooses the good will inevitably come to ruin. With luck a good man might be safe from the many who are not good and prosperous to boot, but one cannot count on such luck. For the good man it is in a strong sense fitting (*conviene*) that he come to ruin, for holding the wrong presumption. He deserves it. Machiavelli does not *expel* fortune but he also does not *suffer* it.[5] Prudence for him is not to take account of risk when necessary but rather to do so in principle, always avoiding evil by presuming that it will be encountered. Thus this passage anticipates his nearly explicit statement that one must do evil to the other fellow before he does it to you (*D* 1.52.1). You may not succeed, to be sure, because the contingency of things may go against you. Perhaps too, the good person will not be punished for his goodness. But with the correct presumption, you stand a better chance.

The presumption of necessity is supported by the impending presence of ruin as the profession of good is not. Who wants to come to ruin? When confronting the stark face of necessity, almost everyone is easily persuaded, or, since persuasion may not be necessary, easily moved toward safety regardless of imaginative persuasions to do otherwise. If necessity is not apparent, it can be made so, and often better with actions than with words. Its being apparent, or easy to make apparent, is part of the simplicity that gives it power and makes its truth "effectual." Necessity has the spontaneity of animal nature behind it, whereas the good needs to be thought about and deliberated. So the presumption of necessity is less presumptuous than the presumption of good. "Nobility" is a delusion that depends on a life beyond life that does not exist; it is an imaginary form of self-preservation. Machiavelli will teach those who desire nobility how truly to attain and assure it.

Here, speaking so emphatically of necessity, Machiavelli takes a long step in the direction of scientific determinism, but he does not go the whole way. By retaining the need for good fortune, he holds to human freedom and virtue in the management of fortune. For Machiavelli, prudence seems to be the same as "shrewdness" (*astuzia*), not distinct from it as with Aristotle.[6] Reason in practice, and so also in theory, is not on the side of goodness.

Machiavelli shows his awareness of the need to go beyond morality in order to question it by speaking of a new sort of truth that will settle the dispute over morality, the "effectual truth" (*verità effettuale*). The effectual truth is opposed to the imagined truth stated in professions of goodness, and it is shown in effects. Near the beginning of the *Discourses on Livy*, Machiavelli shows that the disputes between the nobles and the plebs in the Roman republic should not be condemned, as writers under the influence of classical political science, including Livy, had done. This criticism was based on an imagined harmony between the two typical parties in every republic, but Machiavelli contends that, in their effects, the disputes were the cause of Rome's becoming strong and free (*D* 1.4–6). The "effects" were the outcome in practice, as we would now say, in *effect* or in *fact*, of conflicting opinions that might be resolved on the level of imagined theory, but in the world as it is, would be resolved only as they made men act. In this case the superiority or nobility of the nobles was not deferred to by the plebs but understood as oppressive and opposed, and the result was a contentious republic that had the power to expand and the prudence to satisfy (or at least to appease) the people. In *The Prince*, Machiavelli's discussion of morality after announcing the idea of "effectual truth" explains that the various virtues, called "qualities" in chapter 15, take effect in the ways in which they are "held" (*tenuto*) to be, not as they are. Liberality, for example, is what it will be held to be—its effectual truth—not what it is imagined to be (*P* 16.64). This sort of truth will later be known as empirical because it is based on "fact." To the ancients, a fact was a "that" (*hoti*) to which one could point, but "that" comes and goes, and is not truth, which is permanent. *Facta* (*erga* in Greek) were deeds as opposed to speeches, not truth as opposed to imagination, as for Machiavelli.[7]

Machiavelli's profession of necessity develops a context in which necessity will be understood and appreciated rather than ignored, set aside, or suppressed, as happens with professions of good. This context is the "world," which he constantly invokes, together with the "worldly things" of which it is composed.[8] The world rejects the invisible next world of Christian belief and theology, together with the intelligible world of classical rationalism. The world is visible, and consists of simple and mixed bodies: the simple bodies of nature and the mixed bodies of nature and human forming (*D* 2.5.2). There are no natural forms to be seen, only forms of human conception to be "introduced."[9] The prudence of a prince can put his form on the material of his principality (*P* 26.102), in the political deed that Machiavelli offers to describe human knowing. A prince knows what he is doing when he is introducing his "form," which is making his presence, that is, his truth, effectual. Knowledge of the world is not distinct from acting upon it, for the world's necessities, when understood, open the way for the prince's intervention in it. The neutrality of "worldly things," which are permanent though not intelligible, permits and promotes the enterprises of princes and captains.

As the world of sense, the world is a "whole" on its own. It is not a whole with parts, as it is composed of unintelligible "things" that behave according to necessity. Necessity is divided into necessities, especially in regard to humans, where each human being has his own necessity for which he must exercise his own arms. We are all set against one another in a manner later to be formalized in Thomas Hobbes's state of nature. But there are also groupings of men very relevant to politics, particularly the division of "humor" between those who desire to command other men and those who desire not to be commanded (*P* 9.39; *D* 1.4.1, 1.5.2, 1.7.1; *FH* 3.1). A humor is a medical term that refers to exhalations arising from the body, not the soul, hence indicating a typical necessity rather than a typical choice. Conflict between the two parties of nobles and plebs in the Roman republic made it "free" as well as strong. It was indeed "the first cause" of keeping it free (*D* 1.4.1). Freedom can be found in both the princely and the popular humors—as the freedom to command for princes, ultimately "to be alone" at the top, and as the freedom to oppose or resist being commanded for the people. Each humor, each aspect of freedom, is felt as a necessity that determines behavior rather than a moral choice that guides it.

If the world can be known, and knowledge is of permanent things, does that mean that the world is eternal? This would place Machiavelli in the ambit of Aristotle, for whom the world of joined matter and form is eternal since the natural forms are eternal. But Machiavelli seems to deny that the forms of nature are eternal; rather, there are certain motives or causes of behavior, such as the princely and popular humors. The "simple and mixed bodies" he speaks of, a division of bodies, would appear to signify materialism, suggesting a source in Lucretius, whose poem Machiavelli himself copied by hand. But Lucretius said that the world is not eternal; it is merely a temporary, chance formation of atoms, which are alone eternal. Machiavelli is serious about politics and the knowledge of politics, as Lucretius is not. Perhaps the eternity of the world, inferred so as to make it knowable to himself and later Machiavellians, was accepted by him simply because it was opposed to the creation of the world asserted in Christianity. He was with Aristotle so that the world could be known, and with Lucretius so that he could deny intelligible natures in the world.

Necessity, for Machiavelli, is expressed in the world of sense. Perhaps it is not necessarily expressed there, but Machiavelli inflates necessity beyond its normal confines. In that world, ruin for the body is graver than perfection or salvation for the soul, which does not exist in it. Knowing that world requires learning how not to be good among the many who are not good. This means adopting the goal and practices of acquiring. "And truly it is a very natural and ordinary thing to desire to acquire" (*P* 3.14). Moral condemnation of it is effectual only when one attempts to acquire and fails. One might believe, and Machiavelli at first says, that a hereditary prince, who hardly disturbs the people he rules, is a "natural prince" because he has "less cause and less necessity to offend" (*P* 2.7). But in view of the natural and ordinary necessity to acquire, Machiavelli corrects his view of the natural prince; it is not the hereditary prince but the new prince. On his arrival in power, the new prince cannot help offending both his enemies, whom he has displaced, and his friends, whom he may disappoint (*P* 3.8; cf. *D* 1.7.2). The necessity to acquire applies to the hereditary prince as well, because he must take care to stay ahead of those whose desire to acquire will operate against him. Anticipation becomes the rule of those who hold an acquisition as much as those who seek to gain one (*D* 1.6.4, 1.52.2). The fear of losing generates the same ambition as the desire to acquire, but to greater effect, since the holder of the state has greater means (*D* 1.5.2).

How new must the new prince be? How far does the necessity to acquire extend? It seems at first that the new prince must depend on his "opportunity" to acquire, for example, that Moses found the people of Israel oppressed by the Egyptians (*P* 6.23). But we are told further that a prince can build his own "foundations" so as to make his own opportunity (*P* 6.25, 7.27, 32). Those foundations might consist of the customs and opinions of his time as enshrined in religion, particularly the Christianity Machiavelli found in his own time, which held such little esteem for the "honor of the world" that he thought necessary to defend (*D* 2.2.2). A prince would then have to change the thinking of his people, creating for them a new "sect" and becoming himself the "prophet" of that sect.[10] The best way to do this might be not to create a new sect but to take the existing sect and transform it to one's own purpose. This is what Machiavelli did to Christianity, as he shows by citing and reworking the example of David and Goliath in chapter 13 of *The Prince* (*P* 13.56).

The necessity to acquire compels one to use force and fraud, especially fraud. Those who rise "from small beginnings to sublime ranks," and Machiavelli cites his exemplar the Romans, always find it necessary to use fraud and are "the less worthy of reproach the more it is covert" (*P* 7.32; *D* 2.13.2). Obviously, they cannot use "open force" at first, when they are weak, but they could excuse themselves from blame for using fraud by remembering that the power they displace also rose to its height by the same means. And as with the necessity to acquire in order to maintain, so too with the necessity for fraud: a powerful prince needs to use fraud to protect himself against the fraud that will be used against him. Thus, because acquiring means acquiring from others, or in competition with others, secrecy becomes essential to politics. When anticipation is the rule, one cannot afford to let others, that is, one's enemies, know what is being planned against them. The characteristic mode of behavior becomes conspiracy, not only in politics but in all society influenced by politics, including friendship. It is a "very wise thing" to live a life of conspiracy, and a necessity that must be judged particularly by the few who might prefer the contemplative life.

That religion is a kind of purging is shown by Machiavelli's worldly reduction of it. For him, morality and religion are effectively the same, because morality cannot prove that humans are able to afford to be moral without recourse to divinity in the next world. The god must be there to punish and reward, and to do so he makes commands on humans who, as such, are imperfect sinners. Yet most humans—the people—are imperfect sinners because they are too weak to sin without fearing the consequences. So they must have religion—but they cannot live by it. The necessity of living by religion is counteracted by the necessary impossibility of doing so. Men, being sinners, cannot live without sinning. So they need a church and a priesthood that both enforces the commands of religion and provides the relief of forgiveness from those commands.

For Machiavelli, religion is not the overcoming of the world's necessities that it claims to be. In its promises as well as demands, it accords with the necessity arising from human weakness. It is not concerned with goodness, or not so much as with predicting and controlling the future, hence providing security for human weakness. More than finding remedies for their faults, humans want to know what is in store for them as they are. They prefer security to reform. Religion is essentially an attempt to master fortune, but "the present religion," as he describes Christianity (*D* 1.pr.), does this in the interest of priests who do not believe in it. His own replacement or reformation of Christianity, putting it under the mastery of the princes of the world, does no more disrespect to Christianity than Christianity has done to itself. Christianity, he says, "shows the truth and the true way," a careful statement that falls definitively short of saying that it is true (*D* 2.2.2, 3.1.4, cf. 1.12.2). Christianity will indeed show the truth and the true way as Machiavelli appropriates it to his own use, in accordance with its "effectual truth," as modes and orders of human government with which, as we have seen, he will redeem the sinfulness of the world. His own atheism will take advantage of the atheism of Christian priests who "do not fear the punishment that they do not see and do not believe" (*D* 3.1.4). But it will not be able to abolish religion and will not try.

Religion and pre-Machiavellian morality will continue in the world as the humanly necessary dissatisfaction with its uncertainties and the unpredictability of fortune. Machiavelli surely encourages cynicism about morality, but he knows that he cannot convince most people to abandon morality. Resistance to necessity in the form of morality is as necessary as the failure of morality. "Goodness is not enough" (*la bontà non basta*, *D* 3.30.1), but it will not disappear, and people will not stop thinking that it is enough. Hence Machiavelli does not attempt to construct a universal new morality, as did later thinkers, on the basis of a universal right of self-preservation. According to him, virtue for the princes will always have to contend with goodness for peoples (*D* 1.18.3–4). The prospect that he might become known as Old Nick, and his advice become notorious as "Machiavellian," would neither have deterred nor surprised him.

The Machiavelli scholars who try to save his reputation will succeed with many who are as credulous, and, in their way, as moral, as those scholars are.[11] But Machiavelli himself would have excused them, because they operate on a necessity he understands better than they. It should be noted that Machiavelli does not justify so much as excuse evil.[12] He himself, so to speak, personally "excuses" the homicide of Remus by Romulus as well as the failure of Piero Soderini to anticipate the evil that the Medici did to him (*D* 1.9.2, 5; 1.52.2). The primacy of forgiveness over justice in Machiavelli's thought reveals his desire to replace, and assume the office of, the giver of forgiveness—and betrays the permanent tinge of Christianity to his anti-Christian thought.

A special challenge for the moral scholars is the Machiavellian speech of an unnamed leader of the plebeian Ciompi rebellion in Florence, which, because of its repeated reliance on "necessity," deserves (and bears) close examination here (*FH* 3.13). The orator speaks, it is said, "to inspire the others," but he says that he teaches what necessity requires.[13] Apparently necessity sometimes needs to be inspired in those to whom it applies; necessity does not necessarily make itself effectual but has to be taught in a striking way. He begins by saying that if he had to deliberate whether to take up arms, burn and rob homes, and despoil churches, he would agree "to put quiet poverty ahead of perilous gain." No moral qualm at these deeds would occur to him! But speaking now in the midst of rebellion, he says that we have no choice but to multiply the evils already committed, and add more companions in them so that more will suffer, because universal injuries are borne more patiently than particular ones. Thus can we gain pardon more easily as well as "live with more freedom and more satisfaction than we have in the past." Here is Machiavelli the champion of republican freedom and virtue.

The orator goes on to disparage the nobles who oppose the plebs. Don't be dismayed by their antiquity of blood, he exclaims: "Strip all of us naked, you will see that we are all alike." Forget conscience and possible infamy, for where, as with us, there is "fear of hunger and prison, there cannot and should not be fear of hell." And then he generalizes grandly: "For faithful servants are always servants and good men are always poor." "God and nature" have put us where we are, in the midst of exposure to wickedness.

So, "one should use force whenever the occasion for it is given to us." This is a course of action, the orator confesses, that is bold and dangerous, but when necessity presses, boldness is judged prudence. "Spirited men never take account of the danger in great things, for those enterprises that are begun with danger always end with reward."

The original claims that this circumstance is special, and the initial concession that one should hesitate over "perilous gain," are entirely withdrawn. To be in the midst of a plebeian rebellion is not exceptional but reveals the essential situation of man: all of us stripped naked, exposed to danger and wickedness. Here is Hobbes's state of nature, and not just in embryo but born alive and kicking. The nature of man's situation is taken from the extreme case and made universal to cover all normal cases. In fact, the concept of "normal" as opposed to abnormal is reversed so that the abnormal, formerly the exception, becomes the rule. Here, too, the future course of modern science is previewed: the nature of man is taken from nature stripped of convention, as it were in a laboratory experiment, when nature is tortured and everything normally hidden emerges. In the practice of experiment, scientific fact is disclosed as opposed to ordinary observations made complacently without benefit of the pressure of necessity. We should also notice the ambition of the orator for great things and his willingness to face great danger in enterprises with the expectation of reward.[14]

The plebeian orator who "inspires" the mob by appealing to necessity shows again the unexpected complexity of Machiavelli's profession of that notion. Humans must not only choose necessity but also be inspired to choose it. That politics is ruled by necessity does not at all mean that political things must be accepted as they are with resignation, leading to disdain for the political life and the search for "quiet poverty" in contemplation that, perhaps, the orator momentarily considered attractive.[15] Instead, Machiavelli seeks to inspire (*inanimare)* his readers with a spiritedness (*animo*) that will lead them to virtue (*virtù*) in the sort of active acquisition that he defines as the political life. *Animo* easily recalls the *thumos* by which the classical political scientists referred to the spirited part of the soul. Machiavelli does not mention the soul in *The Prince* or the *Discourses on Livy*, his two chief works, and he seems to treat *animo* as his replacement for soul, substituting *animo* for *anima*.

The necessity for *animo* is further complicated by the complacency of routine that all the achievements of virtue induce. Machiavelli presents this untoward consequence of virtue in a well-known passage in his *Florentine*

Histories discussing the fourth of his seven inquiries in that book (*FH* 5.1). There seems to be a cycle in history in which "virtue gives birth to quiet, quiet to leisure, leisure to disorder, disorder to ruin," and then from ruin to a rise, in reverse order, through the stages. The danger of leisure is paramount, and especially the use of leisure for philosophy illustrated in the protest by Cato against the corrupting presence of philosophy in Rome. Applying this dilemma between what is good for a republic and what is good for philosophy to his time, and to the difference between the "virtue and greatness" of the ancients and the weakness of the moderns, Machiavelli closes with a suggestion. "Perhaps," he says, it may be no less useful to know the modern weakness than the ancient strength, because if the latter excites "the liberal spirits" (*i liberali animi*) to follow, the former will excite such spirits to avoid and eliminate it. That is a statement of Machiavelli's liberal spirit, and apparently, he has a remedy for avoiding the cyclical necessity of virtue and disorder.

The general program for a lasting or even permanent revival out of weakness is supplied in his *Discourses on Livy*. There he concludes that "nothing is more necessary" in any common way of life "than to give back to it the reputation it had in its beginnings" (*D* 3.1.6). Necessity leads out of necessity, when prudently understood as requiring a return to the beginnings and to the original fear that underlies the complacency of civilization. This return has to be a political act, a sensational change of regime that catches attention, as opposed to the steady accumulation of property that later Machiavellians, agreeing with Machiavelli as to the necessity to acquire, substituted for the riskiness of Machiavellian virtue.[16] As virtue is risky, the goal of virtue is glory, which one might say is a semblance of nobility. In the sense of glory, nobility is not opposed to necessity but rather gained through necessity, an insight for which Machiavelli praises "certain moral philosophers" he does not name (*D* 1.4–5, 43; 2.12.3, 3.12.1). Fraud, for example, might seem to be necessary though "detestable"; but no, fraud in managing war is glorious (*D* 3.40.1).[17] Machiavelli's use of fraud, one might propose, is the highest degree of his glory.

So understood, Machiavelli can return his profession of necessity to the profession of good, the latter having been duly limited and disciplined. The new prince must arm his subjects, not all of them because that is not possible, but some of them. Which should he choose? On thinking it through, he will see that it is easier to gain to himself those who had been content with the previous state, his former enemies, rather than with his former friends, who had their own reason for becoming so and would be more demanding of him (*P* 20.83, 86). In Machiavelli's own case, he would, one supposes, be thinking of Christian priests as his new friends. Such a course may not be perfect, but one can never seek to avoid one inconvenience without running into another, and prudence consists in picking the less bad inconvenience as good (*P* 21.91). The next chapter of *The Prince*, on the secretaries to the prince, discusses only one case: that of a minister who is more excellent than the prince he advises. One knows of necessity, Machiavelli says, that this prince was either in the first rank of inventiveness or the second, being able to recognize good deeds though incapable of conceiving them. This minister "cannot hope to deceive him and remains good himself" (*P* 22.92).

At least in the case of a minister advising a prince, then, it is necessary for the minister to appear good and faithful. Machiavelli as adviser to princes is of necessity faithless to any particular one of them, because his advice is general or universal and can be used by the enemies of any prince whom he advises. But Machiavelli is himself also under the necessity of proving to be good for the princes he advises and not merely offering irresponsible advice in order to make himself look good. We see that necessity is judged, in the end, by how much good it leads to—even if the good in this case is only apparent. Machiavelli, the professor of necessity, is obliged to profess the necessity of the good. His early adumbration of the fact/value distinction is obliged to accept the fact of value as well as the value of fact. One gets a glimpse of the humanitarian science in Francis Bacon's thought, powerful today, in which the noble cause of man shines forth from the objectivity of science, oblivious to any cause. Science is a substitute for religion above all in its claim—fraudulent?—to be above man, above his world, comprising his interests and his necessities. It has forsaken Machiavelli's focus on *uno solo*, in favor of being an open conspiracy of adepts and believers without a prince, or without any prince known to it. Instead of the false imagining of revelation, science proposes the authority of fact, evident fact, evidence—the visible. Even the invisible in science leaves its trace in what is visible for the human eye, which however aided by an instrument, still needs to look. With this continuing dependency, science reveals its reliance on nature and thus its inability to master nature by reducing it to mere fact.

Chapter 2

The Place of the Treatment of the Conquest of Nature in Francis Bacon's *On the Wisdom of the Ancients*

Svetozar Y. Minkov

> The question between me and the ancients is not of the virtue of the race, but of the rightness of the way. And, to speak truth, it is to the other but as Palma to Pugnus, part of the same thing, more large.
>
> —Bacon, letter to Mr. Tobey Matthew (1609)

> From that zeal and constancy of mind, which has not waxed old in this design, nor after so many years grown cold or indifferent. I remember that about forty years ago I composed a juvenile work about these things, which with great confidence and a pompous title, I called Temporis Partum Maximum.
>
> —Bacon, letter to Fr. Fulgentio (1625)

Bacon is famous as a proponent of the "mastery of nature," even though he never uses the exact expressions "conquest of nature" or "mastery of nature," just as he is famous for the scientific method despite the fact that "method" was a term of opprobrium for him.[1] He does, however, speak of "command in nature" (*imperium in naturam*) and "winning over nature" (*vincere naturam*). And he does so not in casual asides but, among other places, in two myths—"Erichthonius, or Imposture" (§20) and "Atalanta, or Profit" (§25)—found in the midst of an entire sequence of myths devoted to the theme (§§19–29).[2] That sequence itself is found in *On Wisdom of the Ancients* (1609; henceforth *WA*)—a work about the relative merits and demerits of ancient wisdom.

WA suggests that the mastery project is part of Bacon's larger intervention in the relation between philosophy and society. At its core, this intervention consists in Bacon's manipulation of the newly enfranchised populace—a populace partly empowered by Bacon himself—and his hijacking of their religion. The first nine of the book's thirty-one parts are devoted to an explication of the politics of his strategy. The manipulation takes place through seditious rumors and controlled rebellions. In the case of Bacon himself, there is a corresponding rebellion against the tradition, based on a reinterpretation of the fundamental moral and cosmological beliefs of that tradition. The book also includes an argument that such manipulations and rebellions are not beneath the dignity of the philosopher. It shows that Bacon's project is a philosopher's *instrument.*

Following the architecture of the work, the conquest of nature is to be understood in the light of the four thematic clusters that precede it: (1) the Machiavellian political analysis enriched by natural-scientific and cosmological considerations; (2) the humiliating, if perhaps toughening, cosmology of aimless, repetitive chaos (having stressed this chastening cosmological chaos, Bacon could then offer secular charity as a much longed-for relief: Christian morality without the Christian God, but with the vague promise of the return of the Christian god[3]); (3) the context of Bacon's acute psychology of pity, self-abasement, and devotion to humanitarian ends; and (4) Bacon's stern warning against a frontal assault on religion.

The work consists of thirty-one ancient fables and parables, each followed by Bacon's interpretation, with two dedicatory letters and a preface at the head. Though elusive, on account of its allegorical character (and especially because of Bacon's reticence on the nature of the transition from tale to tale), *WA* contains Bacon's most sustained treatment of the mastery of nature. The theme is treated in a section of eleven fables and parables (§§19–29).[4] If one includes in this count tales on the mastery of human beings, the number rises to twenty-nine, if not thirty-one. One might perhaps exclude the fables that are frankly cosmological, but even these tales belong to the overall project of providing a new theo-cosmological scheme in which science can come forth as the new savior of humanity.

My aim, then, is to understand the significance of the placement of the topic of the conquest of nature within the whole of *WA*. As Bacon indicates in the preface, *WA* is a combination of civil-moral fables (the themes of the *Essays* and the hard political realism of the *History of Henry VII*) and natural-scientific ones (corresponding to Bacon's scientific writings), as well as a combination of the old and the new. It also includes the reflections on the method, primarily in "Scylla & Icarus, or the Middle Way" (§27).[5] Thus the *Novum Organum* as well as *Of the Proficience and Advancement of Learning, Divine and Human* are also compactly represented in *WA*.[6]

The section in *WA* devoted to our theme, "the mastery of nature," is preceded and prepared by two other sections—a political section on using and suppressing popular rebellions (§§2–9) and cosmo-theological section including a crucial digression on pity and Bacon's self-presentation (§§10–17 and 14–16)—and is followed by a discussion of Bacon's grand strategy (§30) and his comprehensive summary of the problem of happiness (§31). In one sense, the sequence on the conquest of nature is the *culmination* of the work; in another sense, it is *qualified* by the preceding parts. Its parable-like character likewise casts a shadow on it. The use of parable for the sections devoted to the conquest of nature suggests that Bacon's project has the character of a quasi-biblical preaching. Now, one might ask, "Isn't calling a story a fable even more insulting than calling it a parable?" And indeed, at the end of one of the three fundamental natural-scientific fables in the work, "Coelum, or Origins," Bacon says, apparently derogating the cosmological account he has just given, that "this fable has a philosophy and this philosophy has a fable"; but in this case he is indicating that his presentation of materialistic cosmology has at least some traditional elements, accommodating the theological view. In the preface to *WA*, Bacon had included "fables, enigmas, parables, and similitudes" as the devices necessary for an easier and more benign approach to the human intellect at a time when the minds of men were still crude, but he hastens to add that this is necessary even in his own time. Perhaps "fable" insinuates gently a deep but harsh teaching while "parable" presents a new edifying doctrine.

* * *

WA begins less unscrupulously, with mere dissembling (though outright simulation is practiced, it is not mentioned by name in the book), with the issue of speaking truth only in the appropriate way, at the appropriate time, to the appropriate audience, taken up in the first fable, "Cassandra, or Outspokenness." It is puzzling that Bacon claims that Cassandra possesses powers of divination and, moreover, is artful and capable of dissimulation and yet fails to understand (and control) her situation. Bacon omits the traditional parts of the fable in which Cassandra goes mad and tries in vain to predict her own painless death, so he could have painted her as even more of a failure. He may take "pity" on Cassandra if he associates himself with Cassandra, since he himself is making wild, "utopian" predictions about the future and is likely to be disbelieved; he did not publish his *New Atlantis* during his lifetime. And he is capable, like her, of "keep[ing] hopes warm." But Cassandra lacks the deeper knowledge about how to provide counsels and warnings at a seasonable time.

The problem of the relation of political power to science is dramatized vividly in "Cyclopes, or the Ministers of Terror" (§3). Jupiter had to resort to proxy "ministers of terror" in order to get rid of Aesculapius. Aesculapius had raised a man from the dead with medicine. Jupiter was angry, but he could not show his indignation, since Aesculapius's deed was considered pious and celebrated. But the political power of Jupiter would have been rendered fundamentally futile if a scientist could come into the possession of a resurrecting power. Jupiter could kill enemies or subordinates, and the scientists could bring them back to life. The fable seems to "pertain to the actions of kings," and it does show the king winning over the scientists. Aesculapius did not have a scientist friend who could resurrect him, nor could he resurrect him after being struck down by the bolts of the Cyclopes.[7] This is the *first* step in Bacon's treatment of the relation of scientific progress or political power. The immediately succeeding fable, "Narcissus, or Self-Love" (§4) suggests that, politically speaking, a philosopher cannot afford to "lie low" in a merely contemplative posture.

It is Bacon's stratagem to insert a digressive section in a unified thematic sequence in order to deepen his investigation and jolt us into juxtaposing distinctive fundamental thoughts. "Pan" (§6) is such an insertion, extraneous to the political-moral Machiavellianism that makes a unity out of fables §§2–9; it is as if Bacon is asking us to think of the cosmological underpinnings and implications of Machiavellian politics. It is a fable placed digressively into the political-moral sequence, §§2–9, most of which concerns "rebellions" (especially the sequence's first and last fables; rebellion is also the theme of the first three fables in the preface). If rebellion is taken allegorically as a rebellion of thought, a deep fable on the fundamental matters is not out of place near the center of the rebellion fables. "Pan" is indeed such a revolutionary analysis of nature, which is not organized purposefully but rather is the result of a materialistic confusion of seeds, and therefore it can be manipulated without fear of disturbing its beauty or order. The person who reaches this realization will be in a position to establish an *imperium* and a find a basis for a harmonization of philosophy with society.

Having shown the relative failure of ancient political and natural philosophy in the image of Actaeon and Pentheus (§10) and the overly erotic Orpheus (§11),[8] Bacon presents two fables—"Coelum, or Origins" and "Proteus, or Matter"—in which he offers a new, stripped-down cosmology as part of his experimental wooing of the vulgar. In "Pan," Bacon had said that there are two fundamental alternatives: either materialistic confusion of seeds or divine word and providence. In "Coelum," contrary to his own admonitions to keep science and religion separate,[9] Bacon says that Democritus came close to the truth of the divine word, said to be the eternity of "unformed matter," which, together with chaos and Cupid (the inner tendency of matter), precedes everything. Accordingly, in the same fable, Bacon traces, through its fibers and to its root, a fundamental theo-cosmological mistake—the false assumption of positing God or soul—which arises from the neglect of the fact that matter is active. When one takes this into account, one is able to explain the visible order on the basis of matter's random, countless agitations and motions which are unintentional "attempts at worlds." For the same reason, the walls of the current world are insecure and can relapse into ancient confusion at any point. This fragility of the world underlies the inescapability of human mortality. Bacon quotes Lucretius merely "praying" that what his reason tells him (that the world will collapse) will not happen in his lifetime, which is to say, will not correspond to how things are.

"Coelum" is followed by "Proteus" (§13), which is about the stretching of nature.[10] In "Proteus," the only one of the thirty-one tales in which he speaks of a "miracle" or of omnipotence, Bacon brazenly cites the human lack of divine omnipotence as the reason humans can try to destroy nature through extreme measures (humans are not God and so cannot succeed; nature will only produce miraculous new shapes under extreme heat). And nature, not God, is the true prophet: Proteus, under investigative pressure, will tell of things past, present, and future.

But before he completes his subversive cosmology in "Cupid, or the Atom," Bacon *interrupts* himself and provides three ostensibly modest or contemptible, but in fact fundamental, moral-psychological fables. Fables

§§14–16 are the second and central digressive interposition—in this case, in the theo-cosmological sequence §§10–17. Bacon thus almost forces the reader to reflect on the political meaning of his new materialistic cosmology. Even beyond that particular feature, it is typical of Bacon in this book to interweave and shuttle between metaphysical and political themes, or conversely, to mix the high and the low, a procedure beautifully exemplified by the central fable (§16) in which the highest takes on the guise of the lowest: the king of the gods masquerades as the miserable, rain-drenched, stupefied, trembling, half-dead, abject cuckoo.

These digressive central fables, §§14–16, nestled between the hard-hitting cosmological ones, are also central to Bacon's project and his psycho-political cosmology. This is the middle instance of three such digressive interpositions: "Pan" is interposed in the political fables and "Dionysus" in the conquest of nature fables. (Interestingly, *WA* itself is a kind of "interjection" in Bacon's oeuvre, not a proper part of the "Great Instauration," as he says in a letter to Fr. Fulgentio.) As he implies in these central fables, Bacon will increase the use of pity and compassion, while simultaneously, in the cosmological fables, leaving out God, or reducing Him to nature.[11] Bacon's practical psychology—his redirection of hope, envy, pride, and admiration—is the meeting point of his cosmology and politics. The "Memnon-problem"[12] of §14, as well the theme of §15—the problem of old age—concerns the excitation of pity for failure and frustrated hope; to some extent, this is also the theme of the failures of Cassandra (§1), Narcissus (§4), and Orpheus (§11), but Bacon means to excite pity for youthful promise cut short or the "indignity" of old age (§15) rather than for the various failures of past philosophic types.[13] As Bacon evokes pity for the noble failures of the human race, he distances the human race from its theo-cosmological comfort and incites noble resolve to do something about it. Bacon tries to found a secular morality: to keep Christian morality while rejecting the Christian God (cf. Nietzsche's *Twilight of the Idols*, chapter 9, section 5). This operation receives its briefest but sharpest expression in the central and key fable of the work, in which Jupiter, the "Prince" and most widely present character in *WA*, is referred to as "Juno's Suitor."

The outright simulation of §16 is prepared by the already noted dissembling of Cassandra (§1) and Endymion (§8). The meaning of Endymion's fable is unfolded in two stages corresponding to two meanings of being a "favorite." In the first, Endymion, who gains the moon's favor, is the subject who is inexplicably the favorite of the prince—a case of the high graciously and inexplicably favoring the low. Insofar as Bacon is the prince, he may be considering how to "fatten up" or keep satisfied (if not stupefied or petrified) the people to whom he is appealing with the technological boons of his new science. But in the second half of the story—in the second and last use of "dissimulation" in the book—Endymion is *pretending* to be innocently and innocuously dull by not asking "forbidden questions."[14]

The final fable of the political section is "The Sisters of the Giants, or Rumor"—the shortest myth in the book but one with the distinction of being discussed in the Preface. This rumor ("*fama*") is not the celebratory fame of Pegasus from "Perseus" (§7), but the "fame" of "seditious rumors." Inciting such "seditious rumors" is Bacon's political mode of operation; he is not a military commander proper, but a "propagandist." He divides and conquers using the "womanly" strategy of spreading rumors—"rumors" about the nature of the universe, the nature of the people, Christianity, ancient philosophy, and God.[15] Like Aristotle, Bacon is a conqueror of opinions and only thereby a conqueror of nations (*AL* II.vii.2: "The one to conquer all opinions, as the other to conquer all nations").

The sequence of fables devoted to *Machtpolitik* (§§2–9, with the important digression of "Pan"), begins and ends with an open vituperation of the nature of the vulgar. In "Typhon" (§2), we hear of the innate depravity and malignant nature of plebs, and in the section's concluding fable, "Sister of the Giants, or Rumor" (§9), the nature of the vulgar is said to be perpetually swollen and malignant toward commanders. When Bacon then moves into the next section of the book, in which he offers his new and tougher cosmology, he eases his disparagement of the vulgar. This is consistent with his wooing of the people. In "Juno" (§16), Bacon does speak of the proud and malignant, but this is attributed to the vulgar *only by implication*—one has to put together §16 with §2 in order to see that Juno signifies the people. Bacon resumes his explicit maligning of the malignancy of the vulgar in "Diomedes, or Zeal" (§18), a fable devoted to religious wars and the inadvisability of a direct attack (by science) on religion. To the vulgar, nothing moderate is ever gracious. "Diomedes" serves as a preface to the conquest of nature sequence (§§19–29), in which Bacon ceases insulting the vulgar, except in the case of "Nemesis, or the Vicissitude of Things" (§22), where the "envious and malignant nature" of the vulgar is mentioned. This is the most Bacon allows himself in attacking frontally a providential punishing god in which the people resentfully have placed their faith. As the preface to the conquest of nature sequence, "Diomedes" suggests that a direct attack on religious passion—an assault that may be moved by excessive anti-theological anger—is an unwise strategy for the philosopher-scientist.

Similarly, there is a digressive tale in the conquest of nature sequence: "Dionysus, or Desire" (§24).[16] What makes this section a clear digression is that it is emphatically *not* about the mastery of the passions. What Bacon tries to achieve is not the control or extirpation of desire but its emancipation. Yet this fable has two very special features. The first is that Bacon strikingly calls it a parable—it is perhaps *the* parable of the book. The second feature concerns the confusion between, or the interchangeability of, Bacchus and Jupiter. Noble, brilliant, and deservedly famous activities sometimes originate in virtue, right, and magnanimity, and sometimes in (ignoble) latent affections and occult desires. But, as Bacon never tires of repeating, *dry light is best;* this is found in "Scylla and Icarus" (§27): as "almost everyone has noted," the soul is "utterly degenerate" when drawing moisture from ground. And Bacchic *furor*—used only in "Actaeon and Pentheus," in "Orpheus," and here in "Dionysus" (the tale which explicitly brings together Actaeon and Orpheus's failures at the hands of religious frenzy)—is the reason for the failure of philosophic types. It is the enemy of "curious inquisition and healthy and free admonition." Dionysus, after all, stands for the merely "apparent good," for fanaticism and superstition, even for ridiculously puffed-up vanity.

The first part in the sequence of conquest of nature stories, "Daedalus, or the Mechanic" (§19), is a cautionary tale about the destructive potential of technology. "Daedalus" is the first story characterized as a parable, inaugurating the only section in which the stories are called "parables," and in which parts of fables are said to be "parables." This peculiarity may be because the conquest of nature is dangerous business and hence didactic, parable-like warnings are in order; or because the conquest of nature itself is a mere parable; or because Bacon is indicating that he is acting as a kind of evangelist. In any case, underneath the parable-like presentation of the story is a hard-nosed reflection on the role of the scientist in society. As a quotation from Tacitus indicates, scientists are always retained and banned; they are always foreigners, to some extent, in their own countries.[17] The Daedalus story also underlines the disjunction between intellectual and moral virtue. Traditional moral virtue gets downgraded so that Bacon can present something ignoble as noble.

Bacon's assessment of the meaning of the conquest of nature is conveyed partly through the placement of the section on that conquest, partly through the density of the word "parable" in the section on that conquest, and partly through its preface and what follows (i.e., "Diomedes" at the front and "Metis" and "Sirens" at the back). Does the late placement of the conquest of nature section indicate that it is the culmination of the work, or does it show instead that it is secondary in importance? Not surprisingly, it is in this sequence of fables that "art" appears with the greatest density. Conquering nature (*vincere naturam*) appears first in "Erichthonius" (§20) and second and last in "Atalanta" (§25). The "command of nature," in explicit juxtaposition with "command over men," appears in "Sphinx, or Science" (§28).

In *the* fable on the conquest of nature, "Atalanta," we observe a race—regulated by certain "conditions," a "law," or a "pact"—between art or "technology" and nature. It is a race to the death of nature, though strikingly not to the death of art. Nature is only saved from "death" because technology is distracted by "golden apples" tossed alongside the racetrack. Ostensibly, these apples are nature's way of distracting humans with temptations and short-term payoffs, away from pursuing a patient and "methodical" technological course. Yet—in the only story in which Bacon refers to philosophy as the "art of living"—Bacon fails to identify nature as the character strewing about the golden apples.[18] Now, Bacon is not averse to the diverting use of gold. In a fable in *NO* I.85, an old man tells his sons that he buried gold in the vineyard but does not know exactly where, so they have to dig everywhere. Could it be that Bacon himself has strewn about three distracting golden apples for his readers?

In "Sphinx, or Science" (§28), science is represented as the murderous Sphinx, because it lends practical urgency to theoretical questions ("Action, Choice, and Decision"). Science is beginning to enter into public life, but in displacing or threatening religion, it unsettles men. That is in part why the Sphinx stands for science, not for nature. In that fable, it is not so much that nature is *mastered*, but that science is *managed*. Once Oedipus solves the *human* riddle, he has *imperium* over humanity. Science—having been "pervulgated" or vulgarized—can be seen as the human benefactress. Oedipus is able to overcome the perception of science as dangerous and as profaning the sacred depth of human life. He is able to solve *and* interpret the riddle because he knows that the most important thing about man is bodily decay and the soul's reaction to it. The need for *tangible*, as distinguished from *promised*, eternal peace and happiness is what man most longs for, and religion is only a promise. Natural science, whose ultimate and proper end is said to be command over "bodies, medicines, mechanics, and infinite others," is, in this sense, more persuasive than religion.

Yet since Bacon omits the crucial and universally known tragic elements of the Oedipus story—the parricide, the incest, the self-blinding, and the transgenerational curse—one can assume that the message is not so rosy. One could argue that it is the elimination of tragedy that is Bacon's contribution, but there are other signs that Bacon knowingly suppresses rather than blithely ignores tragedy. Bacon could hardly have expected the indirect rule over human beings by philosophers through science to continue indefinitely, in part because his own "Sphinx" story suggests that science cannot have genuine command over nature. The Sphinx poses two kinds of riddles: one about the nature of man, and the other about the nature of things. Solving the first kind grants the winner command over man, and solving the second, command over the universe. It is crucial that Oedipus does not solve a riddle of nature—that is, he does not obtain command over nature. Yet the solution to the human riddle brings the demise of the creature that transmitted *both* kinds of riddles.

There is another indication of the "parable-like" character of the mastery project. The conquest of nature is alluded to earlier in "Orpheus" (§11). Bacon speaks there of the "exquisite tempering of nature" rather than of a conquest, which is indeed the more plausible formulation if one can only conquer nature by obeying it. Orpheus's work in the restitution or restoration of the corruptible body is referred to as *longe nobilissimum*, by far the noblest. But the noblest work is, from the point of view of philosophy, a *disgrace* (the high serving the low), as was indicated in the central fable of Juno's Suitor: wooing the people is *maxime ignobile*, most ignoble. Nobility (or the related concept of dignity, which is ascribed to Orpheus in §11) is not the same as goodness or rationality (see *AL* II.xxiii.33—"these grave solemn wits . . . have more dignity than felicity"—and the distinction between happiness and greatness in "Of the True Greatness of Kingdoms and Estates").[19]

* * *

In approaching the final part of *WA*, we may recall Kant's sound instinct in selecting a portion of the penultimate paragraph of the preface to Bacon's

Great Instauration as a motto for the second edition of *Critique of Pure Reason*. While giving an account of his project, Bacon says, "We will be silent about ourselves" (a statement that Kant did not read literally enough), and, in a remarkable protestation, he requests that people "take it as a certainty" that he is not laying the foundation of a sect or a doctrine but of human utility and power. Bacon "protests too much." The "voluptuous" Orphic sect Bacon "plants" is described in the concluding fable-parable of *WA*, "Sirens." Consistent with the nobility of Orpheus's attempted conquest of nature in §11, one finds him resurrected and praising the gods in the concluding section. But the new Orphic religion is the "mastery of nature" represented, in the magisterial conclusion of *WA*, by the philosopher's remedy for the "mediocre and the plebeians": a string of anesthetizing, technological distractions.[20] This culminating fable-parable—in which the various meanings of philosophy are played out—does *not* include the mastery of nature as a fundamental solution to the problem of human life (the relation between truth and happiness). Bacon suggests that the alternatives are, on the one hand, a philosophic investigation of the truth, whether it be ugly or beautiful, accompanied or sustained by extraordinary continence, and, on the other hand, two unphilosophic approaches which need to be combined in order to assemble the "mastery of nature" solution: the waxing up of Odysseus' sailors' ears and the loud singing of the praises of the gods (by a resurrected noble Orpheus, no less), so as to drown out the Sirens' call to each nature to pursue its apparent good.

* * *

Let me summarize the argument and action of *WA*. Part I (§§2–9) has as its effectual truth an instruction in wooing and energizing the vulgar; Part II (§§10–17) is an explanation of how to "disgrace" oneself in ostensibly submitting to the ends of the *demos* while testing the *demos*'s openness to a new and godless cosmology. That is, Part II prepares the people, who are being wooed, for the replacement of divine providence by meaningless chaos while aiming to preserve the "humanitarian" morality of pity or charity, "the corrective spice that admits of no excess." Bacon abases himself in assuming this "aspect as if he pitied man." As for the motive behind this operation, one plausible suggestion is put forward by Leo Strauss: "To make the new science, which destroys scholastic science and the accepted, vulgar beliefs, acceptable to the vulgar, the attack of new science on scholastic science must be presented as an attack in the interest of the vulgar: relief of man's estate (cf. Bacon), new science is the basis of medicine. That is, the new scientist must appear as a charlatan. Descartes just as Bacon appeals to the natural this-worldliness of the man of the street in order to fight the other-worldliness of tradition."[21]

In Part III (§§19–29, §18 being a preface)—that is, in the parables on the conquest of nature—"Nemesis" (§22) has as its subject the limits to the mastery of chance. Initially, the subject is not the problem of brute, impersonal vicissitude or chance but misfortune understood as the result of secret and harsh divine judgment. In the eponymous fable §26, the prototype of the scientist, Prometheus—a symbol both of impiously ingrate "human nature" and of the "author of human nature"—is bound to the "column of necessity"[22] that causes him the deepest pain. Bacon at first encourages a tough ingratitude to nature and God for the sake of improving the human condition, only to emphasize in the last part of the "chapter" that the constancy of soul in reflecting on the "inconstancy of human life" is not intrinsic to the scientific type and indicating Prometheus's lack the toughness to maintain such impious ingratitude. As in the essay "Of Adversity," the scientist has to be redeemed by "Hercules," who sails "the length of the great ocean, in an earthen pot or pitcher; lively describing Christian resolution, that saileth in the frail bark of the flesh, through the waves of the world."[23] Finally, "Sphinx" (§28) concerns the neutralization of the dangers that science poses to the worldview of the vulgar. While in one of the exoteric dedicatory letters (the one to Cambridge as an institution), Bacon writes that when contemplation is transplanted into the field of action, it grows taller and leafier (this much is "certain," but that they go "deeper" is only a "maybe"), in the Sphinx fable he says that when the theoretical riddles are transferred from the Muses (i.e., from theory) to the Sphinx (i.e., to practice), they "begin to become troublesome and cruel." The harmony between philosophy and society may be achieved only through simulation.

* * *

In our time, however, access to philosophy seems obstructed precisely by the twin sisters of "Science" and "History," that is, by the fruits of the success of the Baconian project. In his defense, Bacon might say that the providence represented by Prometheus is a new cave or new idol which is necessarily questionable and will have to be replaced by a new "theophany" down the line. Above all, Bacon has left an escape hatch for the potential philosopher in the creation and transmission of *De Sapientia Veterum*. It is a work that suggests that "Science" and "History" were not meant as a path to philosophy by Bacon but as golden apples to distract the enemies of philosophy.

Bacon's view is that this instrumentalization or politicization of philosophy may *seem* like a popularization or vulgarization—a cheapening of philosophy—but in fact is not, since this process is neither intrinsic to nor expresses or affects Bacon's own innermost thought. Bacon's manipulation or management of worldly affairs is essentially defensive or protective. Bacon "stoops"[24] to the ends of the *demos* or the cave, but he does not understand himself as being abased or as vulgarizing his goal. As he says, in his other defense of philosophy, the *Advancement of Learning*:

> Not that . . . I condemn . . . the application of learned men to men in fortune. For the answer was good that Diogenes made to one that asked him in mockery, "How it came to pass that philosophers were the followers of rich men, and not rich men of philosophers?" He answered soberly, and yet sharply, "Because the one sort knew what they had need of, and the other did not." . . . These and the like, applications, and stooping to points of necessity and convenience, cannot be disallowed; for though they may have some outward baseness, yet in a judgment truly made they are to be accounted submissions to the occasion and not to the person. (*AL* I.iii.10)

Nevertheless, one could still wonder if the administrative, social-scientific, technocratic juggernaut which is also humane, charitable, or philanthropic—the modern locus of authority armed by Bacon—might not generate a kind of superpower backed up by faith in the technical solution of the problem of happiness, a super-technocratic tyrant who will prove impervious to even the most adroit or obsequious flattery by the philosopher. As Nietzsche might say, a problem for a future philosopher.

Chapter 3

Hobbes on Nature and Its Conquest

Devin Stauffer

Some of Hobbes's thoughts on nature and its conquest are much more familiar than others. My aim in this essay is to take a broad look at Hobbes's view in its totality, but I will begin from those thoughts that are well known to many readers of Hobbes. Although it may seem initially as if I am merely rehearsing the obvious, my reason for beginning in this way is that the thoughts in question—which concern the basic practical problem that nature poses for man—are a crucial part of Hobbes's position on the matter under consideration. That they are not the whole of it, however, should become clear as I proceed.

Hobbes's most famous statement on man's natural condition is in chapter 13 of *Leviathan*, the chapter that has come to be known as the "state of nature" chapter, even though its actual title is "Of the Natural Condition of Mankind, As Concerning Their Felicity and Misery." In this chapter, Hobbes advances his signature claim that life in the state of nature is a wretched condition of unrelieved anxiety, misery, and war, in which one must continually struggle for survival against natural deprivations and the dangers posed by other men. If "the life of man" in such a condition is solitary, poor, nasty, and brutish, it is almost a good thing that it is also short (see *Lev.* 13.9); only the supreme importance of mere life, for Hobbes, keeps its brevity from being a virtue.[1] Now, one purpose of this bleak account of our natural condition is to undermine what Hobbes regarded as the comforting but illusory belief that men are cared for by benevolent forces beyond man. Hobbes sought to undermine that belief, not only because it is illusory, but also because it is debilitating, for it tells human beings that they need not be overly concerned with taking steps to protect themselves, watched over as they are by "invisible powers." Hobbes's account of our natural condition, by contrast, is meant to be about as comforting as a slap in the face of a man who has had too much to drink. And it has a similar aim: Hobbes seeks to bring men to their senses, to sober them up, and to make them feel in their bones that the only providence in the universe is that which human beings exercise of their own behalf against nature. Moreover, just as Hobbes sends the message that the only providence is that exercised by human beings, so too he indicates that the only beneficial order is that created by human endeavor. The commonwealth—that is, the man-made Leviathan (see *Lev.*, Introduction, 17.13)—is the only thing that can genuinely protect human beings from violent death at one another's hands, the terrible but predictable end of the lives of those in the state of nature who manage to ward off other natural dangers long enough to become a threat to other human beings. In short, nature is the problem, and man's conquest of it through the construction of the mighty Leviathan is the solution.

This simple picture is what comes immediately to mind for many people when they think of Hobbes's basic position on nature and the human condition. There are, however, several complications that must be taken into account to appreciate the arguments that lie behind Hobbes's position. The first complication is that the problem that nature poses for men is a problem that lies, in large part, *within* men, because it is a problem of men's passions, which Hobbes argues are not naturally directed towards peace and community. In Hobbes's view, it was one of the great mistakes of earlier moral philosophers, especially those who took their bearings from Aristotle, to assume that men are naturally political in the sense of being naturally suited for society. Men, according to an assertion Hobbes makes even more bluntly in *De Cive* than in *Leviathan*, are naturally selfish and concerned above all with their own advantage and honor (see *De Cive* 1.2). As others have pointed out, Hobbes treats the latter of these two concerns as the more problematic, both because honor is a good that cannot be shared and because at least some men have an insatiable desire to exalt themselves by lording their superior power over others.[2] Along the same lines, although Hobbes contends in *Leviathan* that there are in the nature of man "three principal causes of quarrel"—competition, diffidence, and glory (see 13.6)—he does not describe them as equally pernicious. The first two—that is, each man's desire to secure nature's meager resources for himself (competition) and his fear of rivals (diffidence)—can be channeled in peaceful directions. The same cannot be said of the desire for glory, which in some men becomes a pathology that leads them to value triumph over security (see *Lev.* 11.1–4; *De Cive* 1.2–4; *Elem.* 9.1). By Hobbes's account, then, some men are even less suited for society than others, and these lupine men, like nature's Romans, render peace as impossible for men in the state of nature as it was for Rome's neighbors during the period of her expansion (see *De Cive*, Ep. Ded.). It is tempting to say that at least these most irredeemable of men, if not all men, are evil by nature in Hobbes's view. That temptation would seem to have to be resisted in light of Hobbes's contention that evil and sin are meaningless notions in the state of nature (*Lev.* 13.10, 13.13; *De Cive*, Pref., 1.10). But if he denies that any man is evil by nature, Hobbes certainly stresses that all men are dangerous, and he indicates that he regards some as more culpably so than others (see, e.g., *De Cive* 1.4; *Elem.* 16.10, 19.2). It especially with such men in mind that he declares that, outside of civil society, man is a wolf to man (*De Cive*, Ep. Ded.).

Of course, as this sketch has already begun to indicate, the natural enmity among men rooted in their natural passions is not Hobbes's last word on the necessary relationship among men. The second complication in Hobbes's position is that, even as our natural passions are regarded as the source of the problem, they are also seen as the source of the solution. At any rate, they are *one* source of the solution in Hobbes's view. After vividly describing the evils of the state of nature, Hobbes concludes his account in this way: "And thus much for the ill condition which man by mere nature is actually placed in, though with a possibility to come out of it, consisting partly in the passions, partly in his reason" (*Lev.* 13.13). On the heels of this remark—and by way of transition to his account of the laws of nature—Hobbes then declares, "The passions that incline men to peace are fear of death, desire of such things as are necessary to commodious living, and a hope by their industry to obtain them" (*Lev.* 13.14). That Hobbes regarded fear above all as "the passion to be reckoned on" (*Lev.* 14.31) is well known, as is his view that the laws of nature emerge through the cooperation of the passions and reason, with reason seeking out the necessary means to the end to which fear and the other peace-seeking passions incline us (see *Lev.* 13.14, 15.40–41; *De Cive* 3.31–33; *Elem.* 15.1). It is not necessary for present purposes to examine these aspects of Hobbes's account in detail. More important is the broader point that Hobbes's account of the state of nature is not meant merely as an articulation of a problem, and certainly not as a counsel of despair. Far from it: Hobbes intends to give men a reason for hope, and to spur them on to the rational efforts that can remedy their condition. It is not simply true, then, that Hobbes describes nature as brutal and harsh. For at least it must be said of nature that she does not leave men with any problems that she does not also equip them to solve.

This further step in Hobbes's argument raises a question, however. If men are well equipped by nature to solve the problem that nature poses, why did they not solve it long before Hobbes ever wrote a word? That Hobbes thinks they did fail to solve it—at least adequately—is indicated sufficiently by his severe criticism of all earlier moral philosophers, and by his bold proclamation that he is the first to raise political philosophy to the rank of a genuine science that can benefit mankind (see *De Corp.*, Ep. Ded.; *De Cive*, Ep. Ded., Preface). Human beings, Hobbes contends, have never seen their way to the political solution they so desperately need. They have built commonwealths, of course, but never on the firm foundation of a truly rational doctrine of morality and politics (see *Elem.*, Ep. Ded., 1.1; *De Cive*, Ep. Ded.; *Lev.* 31.41). If one reason for this failure is that they have had worse than useless guides in the pre-Hobbesian moral philosophers, that is not the only reason. In Hobbes's view, other problematic forces have intervened, forces that have exacerbated men's most pernicious passions and inflated certain hopes and fears beyond their reasonable and safe measures. In fact, the earlier moral philosophers could not have done as much damage as they did had they not nourished a more basic tendency of men toward dangerous delusions (consider *Lev.* 2.8–9, 12.1–6, 29.15, 44.2–3, 46.18). This is the primary reason that Hobbes intended for his sobering teaching about the state of nature to be disenchanting and to bring men to their senses. For he thought that only if men could be freed from the power of crippling delusions would their natural passions lead them to do the constructive work necessary to create enduring commonwealths. But precisely if that is the case, and precisely if it is natural for men to begin from delusion rather than clarity, then Hobbes must acknowledge that what he is calling for is an unnatural act, a break with the natural course of things. It suffices here to think of Hobbes's account of how prevalent and problematic superstition is among men, especially but not only in primitive societies. If superstition is natural to man but also an obstacle to a rational solution to the human problem, then what Hobbes calls for cannot be described as a simple activation of a natural potentiality. He calls, rather, for the creation of something new, something that requires enlightenment and enterprise as well as the cooperation of certain passions and reason. He calls men, in other words, to a project that is more *rational* than *natural.*

* * *

To this point, many readers will likely find nothing too surprising in my description of Hobbes's position. My account thus far is also likely to be uncontroversial, at least for those readers who do not think that Hobbes was sincere in his declarations that he was a believing Christian. This last qualification must be added for two reasons: there are some scholars of Hobbes who *do* accept his explicit claims about his own faith, and my foregoing account rests on the assumption that they are mistaken. (To give just one indication of the latter point, I have suggested that Hobbes believes, and wants to convince others to believe, that human enterprise, not guilt or repentance, is the fitting reaction to the harsh natural condition in which man finds himself—and it may be added here that he says not a word about that condition being a divine punishment for human sin.) Now, I will not attempt here to marshal further evidence that Hobbes was so far from being a sincere Christian that he was in fact an atheist. Although some aspects of my account are already suggestive of that conclusion, admittedly none of them is decisive; but it would take me too far afield to try in this brief essay to address adequately the much-debated issue of whether Hobbes's professions of faith were genuine. Instead, let me focus on another issue that arises for those who regard Hobbes as an atheist. For, at some point, such students of Hobbes must ask about the basis of his (supposed) atheism, which, given that his atheism is an essential premise of the position I have already sketched, is at the root of his view that men must take matters into their own hands if their condition is to be made into something livable. Let me also venture here the broader suggestion that, not only for Hobbes, but for the early modern development more broadly, the relationship between the emerging call to conquer nature and the rise of modern atheism was not one of accidental coincidence—just as, on the other side, it was not mere whim that led the biblical God to disapprove of the construction of Babel.

But what of Hobbes's atheism? What is its basis? The answer that first suggests itself is the mechanistic materialism of his natural philosophy. The first principles of Hobbes's natural philosophy are that the only true beings in the world are bodies, and that all motion and change arise from the collisions and constant rearrangement of these bodies. According to Hobbes, the very notion of an incorporeal substance is an absurdity, and there can be no cause of motion for any given body other than the motions of a body that is contiguous with it and itself moved by another body (see *De Corp.* 3.4, 9.6–7; *Lev.* 4.21, 34.2, 46.15–21; *EW* VII, 85–86). Now, it is not too hard to see how this conception of reality, or of the universe (see *Lev.* 34.2, 46.15), undergirds the outlook and reinforces the message sketched in the first section of this essay. The universe as Hobbes describes it, consisting of nothing but bodies and their continual motions, is not such that it ought to call forth our reverence or gratitude. And just as the fitting response to the dangers we pose to one another is to try to find a way to overcome them, the fitting response to a silent universe of matter in motion would seem to be, perhaps to feel a certain dread before it, but certainly to get to work controlling it as best we can. Once again, then, Hobbes's doctrines are meant to be at once sobering and inspiring, at once a dose of bitter disenchantment and an emboldening call to enterprise.

It is true that there is a certain difficulty with this suggestion as it applies to Hobbes's natural philosophy, that is, with the thought that Hobbes's mechanistic materialism sends the message of a call to arms for human beings. For although such a vision of reality exalts man as the most advanced being in the universe, and the only being capable of using his inventive powers to transform his condition, it also reduces him to nothing more than a body, or a temporary arrangement of smaller bodies, as much determined by forces beyond his control as any other. Are not even man's constructive efforts on his own behalf dictated by complex chains of mechanistic causation?

Hobbes's answer to this question, to remain consistent with his principles, must be yes. Yet that need not simply negate the positive message that man can take matters into his own hands and act to conquer nature. For it is possible to conclude that, at a certain epoch in the history of the universe, for fundamentally accidental reasons, certain combinations of bodies arranged themselves in such a way as to allow, not only for man's emergence, but also for his development and even his ingenuity. That is not to say that free will emerged, nor that the emergent property of certain combinations of bodies that is man's capacity for inventive thinking entails a radical freedom from nature; but it is to say that man's inventive capacity—his capacity to think of what he can do with something once he has it (see *Lev.* 3.5)—remains undeniable and gives man the ability to make a world within the world, even if, in man's construction of that world, nature is in a paradoxical sense conquering itself. Rousseau argues that man can be thought of as perfectible, even if he is not free.[3] Hobbes would agree, although he understands the meaning and vector of man's perfectibility quite differently.

If it is possible in this way to reconcile Hobbes's call for man to conquer nature with the determinism of his mechanistic materialism, there is a more basic, but ultimately deeper, difficulty that besets his natural philosophy. Simply put: how certain was Hobbes—how certain did he think he had a right to be—that his materialistic account of the universe is true? Admittedly, it may seem surprising to raise this question, since Hobbes presents his view with supreme confidence in some passages. Consider, for example, this remarkably bold declaration from chapter 46 of *Leviathan*:

> The world (I mean not the Earth only, that denominates the lovers of it *worldly men*, but the *universe*, that is, the whole mass of all things that are) is corporeal, that is to say, body, and hath the dimensions of magnitude, namely, length, breadth, and depth; also, every part of body is likewise body, and hath the like dimensions; and consequently every part of the universe is body; and that which is not body is no part of the universe. And because the universe is all, that which is no part of it is *nothing*, and consequently *nowhere*. (46.15)

Can the man who wrote this passage, and others like it (see, e.g., *Lev.* 34.2; *De Corp.* 8.20), really have had any doubts about the conception of the universe it so starkly expresses? Again, Hobbes regarded the very notion of incorporeal substances as an absurdity, and he seems to have been firm in his conviction that the only intelligible view of substances is that they must be bodies. He also expresses his basic axioms of motion—the central principles of his mechanism—with just as much confidence as he does his central claim about bodies, the core of his materialism (see, e.g., *EW* VII, 85–86; *De Corp.* 6.5, 9.6–9.10; *OL* V, 217).

Yet Hobbes's most confident proclamations do not reveal his full thought, because he was also aware of a difficulty that arises from the fact that human beings do not have direct access to the bodies that underlie and cause our experience. Hobbes himself stresses, after all, that our perceptions are only of "phantasms" or "fancies" that arise as mere offshoots of the collisions between unknown and unknowable bodies (see *De Corp.* 25.1–2; *Elem.* 2.7–9;

Lev. 1.4). The world given to us through sense perception—which Hobbes regarded as the necessary starting point of all thought—is a compilation of epiphenomena caused by, but hardly identical to or reliably representative of, the more substantial world of bodies in motion. When thought through, however, this situation—our situation—renders it questionable or unknowable, not only what the true characteristics of the underlying bodies are, but even whether or in what way they may intelligibly be called "bodies." For Hobbes at any rate, it proves to be much harder to say even what a body is than one would expect it to be for so confident a materialist as he appears to be. In fact, it turns out that the notion of body, which is the centerpiece of Hobbes's natural philosophy, is a kind of black box whose contents Hobbes struggles to define adequately. Although he defines a body as "that which, without any dependence on our thought, is coincident or coextended with some part of space" (*De Corp.* 8.1; see also 8.20, 8.23; *Anti-White* 3.2, 4.3), this definition is not entirely satisfactory, even by Hobbes's own lights. For it does not succeed in doing what an adequate definition should do, namely, "by a speech as brief as possible," raise in the mind of the listeners "a perfect and clear idea or conception of the things named" (*De Corp.* 6.13). There are two problems with Hobbes's quasi-Cartesian definition.[4] First, there is a gap between extension understood as the real magnitude of a body, which, like the body itself, exists apart from our thought of it, and extension understood as our mental conception of the place occupied by the body, which is a mere image in our minds. If the former is "true extension" (*extensio vera*) and the latter "feigned extension" (*extensio ficta*), it would seem to be only "feigned extension" that allows us to place the body in imaginary space and thus to form a conception of it as extended, whereas it is only "true extension" that exists outside of our minds (*De Corp* 8.5; see also 8.1–4, 8.8, 7.2–3; *Anti-White* 3.1–2). Second, even in the case of true extension, Hobbes resists a simple equation of body and extension or a reduction of body to extension. Extension is only one accident of any body, and thus a body "is not extension, but a thing extended" (*De Corp.* 8.5; see also 8.2–3. 5.3; *Lev.* 5.10). Similarly, since magnitude is the same as extension, it can be said that a body "has magnitude," but "not that it is magnitude itself" (Latin *Lev.*, Appendix, 1185; see also *De Corp.* 8.15, 15.1; *EW* VII, 227). We remain in the dark, then, about the nature of bodies, and that means that we do not really know what lurks behind or beneath the world of our experience, as the true material and efficient causes of the phenomena we perceive.

Now, it will be said—or, at any rate, could be said—that Hobbes thought that, despite its partial obscurity, the answer "bodies" remains the most plausible answer to the question of the identity of the fundamental constituents of reality and the deepest causes. That, I believe, is true—he did think that. Yet even so, the plausibility and semi-intelligibility of this answer does not remove entirely the question mark hovering over the bodies. And more important, the lack of perfect clarity of Hobbes's own answer—of the answer that he himself accepts—opens the door even wider than it would otherwise be open to the radical alternative that some kind of divine will, rather than the aimless motion of bodies, is in fact the deepest source of all being and change. It is partly in recognition of the fact that he cannot rule out this alternative, I believe, that Hobbes takes a set of interrelated steps in his natural philosophy. First, he acknowledges that the causal explanations offered by a mechanistic physics can never be more than hypothetical (see *De Corp.* 25.1, 30.15; see also *De Hom.* 10.5; *EW* VII, 3, 184). If one reason for this is that there are severe limits on what we can know of what we cannot perceive (see *De Hom.* 10.5; *De Corp.* 6.2, 26.1, 27.1; *EW* VII, 78), another is that it cannot be ruled out that some kind of divine will can sever the mechanistic chains of causality. Second, Hobbes acknowledges that, although his mechanistic natural philosophy rests on the plausible assumption that nothing can move itself or be moved except by that which is already moved, it is impossible to exclude the possibility that the world was created *ex nihilo* and thus, again, that (further) miracles can occur (see *De Corp.* 26.1–3 with 26.7, 27.1; *Anti-White* 27.1, 27.22; *EW* V, 176; *De Hom.* 1.1, 10.5, Latin *Lev.*, Appendix, 1171). Third and finally, Hobbes suggests that it is not essential to natural philosophy, after all, to be certain of the existence of underlying bodies as the causes of the phenomena we experience, because it is possible to move forward *as if it were the case*—that is, on the mere assumption—that the phenomena we perceive are caused by bodies, and to rely on such concepts as we ourselves make or on principles that are admittedly human constructions (see *De Corp.* 7.1, 24.8. 25.1, 26.5; *EW* VII, 183–185). To the extent that Hobbes acknowledges that the key definitions in *De Corpore*, in particular, do not depend on the actual existence of any bodies, since he is merely constructing a conception of nature according to which the bodies and their motions are *regarded as* the fundamental reality, it would seem that he retreats from his most robust and confident statements about the nature of the universe to something more skeptical, tentative, and suppositional. Leo

Strauss captured this retreat with the memorable formulation: "Hobbes had the earnest desire to be a 'metaphysical' materialist. But he was forced to rest satisfied with a 'methodical' materialism."[5]

Now, whether or in what way Hobbes did "rest satisfied" with a merely "methodical" (i.e., methodological) materialism can be questioned. For, as already noted, he certainly at times expresses himself as a robust metaphysical materialist. It is probably more accurate to say, then, that he wavers over the matter: as certain difficulties press upon his thinking, he retreats to a methodological materialism; but the position from which he is retreating remains his core conviction, and that conviction is on less ambiguous display at other moments, when he is not confronting the problems that drive him in a more skeptical direction. More important than his wavering—and a key source of it—is that it would not have been possible for Hobbes to rest satisfied with a merely methodological materialism, at least if "resting satisfied" is taken to mean regarding it as satisfactory. Hobbes must in fact have been quite dissatisfied with any approach that entails a renunciation of the hope that natural philosophy can decisively settle the most important of all metaphysical questions, and thus leaves it an open matter whether we live in a universe that really does consist of nothing but bodies and their determined motions.

Still, dissatisfied as he must have been with a natural philosophy that cannot answer the question of greatest concern to him, Hobbes had to face the fact that he did not think it was possible to do better. In some sense, then, he did rest satisfied—or dissatisfied—with a merely methodological materialism. By way of conclusion, let me consider a few of the consequences of Hobbes's less-than-satisfactory conception of natural philosophy and of the physics he thought possible on its basis. This consideration brings us back to the main theme of this essay, because the most important consequence is that Hobbes moves toward a more practical or utilitarian conception of the aim of science, a conception according to which science is directed more toward the conquest of nature than its understanding. As he puts it in a famous remark in the first chapter of *De Corpore*, "*scientia propter potentiam*" ("science is for the sake of power"). I have already mentioned one reason for this movement: if nature is regarded as a meaningless flux of moving bodies, one has no reason to contemplate it with reverence and every reason to get to work transforming it for the relief of man's miserable estate (see above). Hobbes joins other early moderns, such as Bacon and Descartes, in calling for science to be put into the service of augmenting man's power over nature.

But this much follows—for Hobbes, at any rate—merely from his mechanistic materialism; it does not depend on his skepticism. Hobbes's skepticism too, however, plays a role in impelling him toward a more utilitarian conception of science. For the more one comes to accept that there are severe limits on what can be known of nature, the more difficult it is for knowledge for its own sake to remain the end of science. In this sense, theory's loss becomes practice's gain, especially because the limits of our knowledge of nature are not such as to prevent us from making great strides in controlling it. Indeed, because man's neediness is a problem that can be at least ameliorated, if not perfectly solved, it is in some sense even a welcome thing that the concern to address that problem is ready to step in, so to speak, to fill the space vacated by the disappointed desire to know.

Scientia propter potentiam, however, is not Hobbes's last word on the end of science, or at least not the whole of it. For theory, too, has something to gain from the advance of the new physics, even if it rests on a merely methodological materialism. After all, the progressive conquest of nature promises to do more than to provide men with benefits and relieve their suffering; it promises also to give philosophy or science itself, if not the certainty it seeks, then at least ever-increasing confidence that the central principles of mechanistic materialism are sound. Hobbes envisioned a great civilizational advancement, and not only does the new physics have a crucial place in that vision, but it dovetails in its aims with his political philosophy: both are aspects of the project by which man can use his ingenuity to exert an unprecedented control over his condition. The allure of this vision is theoretical as well as practical, for it promises, with each new advance, to make the doubts about the fundamental assumptions on which the project rests fade ever further into a dark and musty past that has been left behind by the progress of enlightenment and civilization.

The allure of this vision does not mean, however, that, for Hobbes himself, as he contemplated such as prospect, the doubts could ever fully disappear. Nor does it mean that, for those of us who live in the wake of the early modern attempt to remake the world, and for whom at least the most important doubts have faded further, the transformation in our consciousness should be seen as an advance in all respects. For the doubts rested on an awareness of the most challenging alternative to what would become the modern scientific outlook, and thus their fading marks the loss of that awareness.

Chapter 4

Devising Nature: An Essay on Descartes's *Discourse on Method*

Stuart D. Warner

> *Renatus Cartesius*—ille est, cui, transplantatis nominis elementis, acclamare licet: *Tu scis res naturæ.*
>
> —Étienne Chauvin, *Lexicon Philosophicum* (1692)

> It is so difficult to find the *beginning.* Or better: it is difficult to begin at the beginning. And not to try to go further back.
>
> —Ludwig Wittgenstein, *On Certainty*, §471

Discourse on the Method for Conducting One's Reason Well and for Seeking Truth in the Sciences is a long title for a short book—seventy-six pages in the large typeface of the original French (fifty pages or so in most contemporary editions) and a scant sixty-five paragraphs in total, all divided carefully into six parts.[1]

Despite its brevity, the book begins with an untitled "Synopsis,"[2] wherein the content of each part is briefly and simply summarized. Order appears to reign. However, when we compare the summary statements about each of the six parts (as found in the "Synopsis") with the actual substance of those parts, we cannot help but notice myriad discrepancies between what they are said to be about and what we indeed discover. Yet these discrepancies barely get at just how strange the "Synopsis" is. Consider its opening line, where prospective readers are advised that "if this discourse seems too long to be read all at one time, it can be divided it into six parts."[3] There are a number of reasons to be puzzled about this, not the least of which concerns any suggestion that such a brief book might be too long. Irrespective of length, however, Descartes himself has already divided the *Discourse* into six parts: readers of the work truly have no choice about the division that lies ahead of them.

More curious still, and of greater consequence, is that, while over the course of the six parts of the *Discourse*, Descartes uses *je* and its cognates no fewer than 864 times—prompting the judgment that the book is autobiographical in character[4]—the first person singular appears in the "Synopsis" not at all. Instead, reference is made to "the author" and what "he" is seeking and trying to accomplish. The oddity of using the third person signals that even though the "Synopsis" has Descartes's book as its subject, Descartes is not being represented as the author of the "Synopsis." Rather, Descartes has ironically distanced himself from the antechamber of the *Discourse*, tacitly invoking a literary *persona* in his stead. This fictive *persona* has read the *Discourse* and now presents his own summary of its various parts to the reader—a summary that encapsulates a quotidian understanding of the book, consistent with the received understanding of the seventeenth-century European world of Christendom; a summary that does nothing to prepare the reader for understanding the antinomian character of the book that follows.

From beginning to end, though, every line of inquiry that the *Discourse* follows is conditioned by Descartes's use of a mode of presentation in which irony, misdirection, ambiguity, and discrepancy are constantly in use, and the surface meaning of various passages is simultaneously transparent and opaque, often with the effect of highlighting the hold that common opinion, custom, and convention have on the institutions and fabric of life *inter homines*. A compelling example of Descartes's literary art in the *Discourse* surfaces in the penultimate paragraph of Part I, where, following his lengthy description and evaluation of the education he received at the feet of his "masters" (I.6) at La Flèche, Descartes informs us that, having escaped from the subjection of his "preceptors," and having set aside the study of letters, he resolved "no longer to seek any other knowledge than that which could be found within myself, or else in the great book of the world" (I.14). Descartes then spends the rest of that lengthy paragraph discussing those travels. As he turns to the final paragraph of Part I, he remarks upon the rather "extravagant and ridiculous" customs practiced by the great peoples he visited,[5] the recognition of which prepared him to appreciate the relative ease by which we are governed by custom and opinion, and he was ultimately able to profit thereby by gradually "delivering" himself from such dependence.[6] Having devoted a fair amount of time "to studying the book of the world . . . I made the resolution to study within myself . . . and to employ all the powers of my mind in choosing the paths that I should follow" (I.15). Descartes's initial characterization of "the great book of the world" gives way to that of "the book of the world": the human realm he encountered proved to be not so great, and the knowledge he sought was not forthcoming.

Since travel had not put him on the right path, Descartes resolved, here at the end of Part I, to study within himself. However, he had already made that very resolution at the beginning of the penultimate paragraph; and indeed that resolution preceded the one to travel in the great book of the world. So, Descartes presents a two-fold resolution, which he sets in contradistinction to the study of letters; but then he discusses only the second element of that resolution, before finally turning back to the first (as if he had not made any prior resolution at all). The action of the *Discourse* here, rather than its speech, directs us to the rudiments of an argument to the effect that seeking knowledge within oneself—whether this means the pursuit of self-knowledge or something else—is possible only on the condition of first understanding the dominion that custom, convention, or common opinion exercises over us. The recognition of the cave, the pre-philosophic condition, precedes philosophy.

From at least the time of Chrysostom and Augustine, and certainly up until the beginning of the eighteenth century, it was not unusual for various thinkers to make reference to two different books with respect to God: the Book of Scripture and the Book of Nature.[7] These were the "books" in which God's teaching and God's making were revealed and made manifest. In setting aside the study of letters generally, Descartes effectively set aside the study of Scripture.[8] More importantly, in this context, is that Descartes rather self-consciously eschews any reference to the ubiquitously used expression "the Book of Nature," which has been inscribed by God, and instead gives birth to that of "the book of the world," which has been inscribed by human artifice. Of course, *du monde* rather than *de la nature* naturally suggests the world of society rather than the world of nature; however, in fixing on the former, Descartes is able to suppress here the question of the relationship between nature and God, while foregrounding the inexorable, yet problematic nature of custom and convention.

Descartes's focus on custom and convention intensifies in Parts II and III of the *Discourse*. The opening paragraph of the former famously begins with Descartes in Germany, in November 1619, during what later came to be called the Thirty Years War, a political-theological conflict that left millions dead.[9] While returning from the coronation of Ferdinand II, the beginning of winter detained him, and he sought shelter in the heat of a *poêle*, where, all day, without any conversation and, due to good fortune, untroubled by cares and passions, he had the leisure to converse with and to take hold of his own thoughts. One of the first of these was that "often there is not as much perfection in works composed of many pieces and made by the hands of diverse masters, than those on which one alone has labored." Descartes, then, exemplified this thought in five different ways, beginning with the work of a single architect as compared to that of many architects: "Thus, one sees that the buildings that a single architect has undertaken and completed are customarily more beautiful and better ordered than those that many have tried to refashion, making use of old walls that had been built for other ends" (II.1). Descartes built on this example and turned to cities, comparing those that grow and develop over time to "fortified cities" (*places régulières*) planned by a single engineer. He then moved on to three interrelated instances of law—fixing first on a prudent legislator, who convened with a people from the moment they were first assembled (comparing that to situations in which the conflicts or quarrels of semi-civilized peoples gave rise to disparate laws that served as responses to the contingent difficulties faced); then on God, who has set down the "ordinances" of the "true religion," a religion that would be better "regulated" than the others; and third, "to speak of human things," on the founder of Sparta, whose "laws all tended to the same end." Descartes turned next to the "the man of good sense," comparing his "simple reasonings" (II.1) about the things that present themselves to us with the probable reasonings contained in books of science that are enlarged, little by little, on the basis of the opinions of many individuals. Finally, Descartes concluded with a man guided by reason alone since birth, rather than by the clash of his appetites and the teachings of his preceptors.[10] Of course, what is clear with respect to this last example is that such a man is impossible and thus, at best, an idealization. But this should make us wonder whether all of the other individual masters in this opening paragraph are, to some extent, idealizations as well, never fully realizable in fact, and, if they are, how they bear on Descartes's overall enterprise in the *Discourse*.[11]

The first paragraph of Part II, at which we have only barely glanced, is mirrored in instructive ways by the ones that follow, where the dominant image is that of tearing down buildings: "It is true that we do not see one raze to the ground all the houses of a city for the sole purpose of redoing them in another fashion, and rendering the streets more beautiful; but one very well sees many who tear down their own houses in order to rebuild them" (II.2). As the structure of the second paragraph reveals itself, it becomes clear that the houses and roads of a city serve as an image for the state and other institutions contained therein, while an individual's house stands for someone's opinions. Although it would seem implausible to try to tear down those various public bodies, that is, to reform them by changing everything from the foundations on up, it would be possible—and perhaps perfectly sensible—for Descartes to reform his own opinions in just this manner, and thereby conduct his life better than by merely relying on old foundations. Thus, he endorses his own private reform at the same time he expresses trepidation and caution about public reform.[12] But he also cautions others about ridding themselves of their opinions, since "the world consists almost entirely" of two types of minds—one haughty, thinking too highly of itself and judging precipitously, the other modest, realizing its ability to distinguish true from false is limited, and needing to follow the judgment of others—and neither of these two types is equipped to follow Descartes's chosen path. Had Descartes "but a single master" (II.3), or had the learned conformed in their opinions, he would have fallen, he tells us, into the class of the modest, but neither of these conditions obtained. Descartes thus must be his own master and walk alone, even if in "darkness" (II.5).

It is in the second and third paragraphs of Part II that Descartes begins to initiate his readers into a world of methodic doubt, under the horizon of which he will call into question all of the opinions he had believed credible,[13] either to replace them with other opinions, or to hold onto them if they satisfy the standards of reason. The whole process involves tearing down what is already there, and then rebuilding, rather than constructing anew. However, the second and third paragraphs draw much from the first: the houses and streets of the city; the dwellings of individuals; the state (which in part appears to align with law and religion in the original third examples); scientific education; and the sciences themselves. There are, however, some elements conspicuously absent, the most notable of which is God. But more strikingly, with the possible exception of Descartes himself, there is no single master present: the second and third paragraphs operate in the realm of the many, hence the emphasis in these paragraphs on first tearing down, all directed toward private reformation, in contrast with the stress in the opening paragraph of Part II on constructing from the ground up seemingly *ab initio*. In that opening paragraph, Descartes indicates that many masters are often inferior to one master acting alone, but he does not explain how or why. Indeed, in the example involving architecture, rather than showing one architect working with the same materials as the many, and producing a more beautiful and better ordered building, Descartes, by suppressing any mention of the materials with which the single architect is working, appears to imply that he is working with a completely different set of materials—thus making it appear as if the example does not exemplify at all the thought with which Part II opens.

Having reached this point, though, we are in a better position to infer the intention of Descartes's initial thought about the one and the many with respect to masters and perfection. The problem with many masters is not that they are many *simpliciter*; that is, suffering from some version of too many cooks spoiling the broth. In fact, as Part VI of the *Discourse* makes clear, the enterprise of science requires an intellectual division of labor, a collaborative effort with many working together in order to reveal the truth of things.[14] The problem with the many, Descartes seems to be suggesting, is that, being many, they are more likely to fall in line on the path of custom, convention, or common opinion, and it is this, being tethered to what is "old," that is the root of their inadequacy, compared to the work of a single master, who is less likely to be similarly afflicted.

It is unquestionable that the example from the opening paragraph of Part II that comes closest to Descartes himself is the man of pure reason alone, which, as we have noted, is surely a practical impossibility. Such a person, though, if possible, would be marked by the capacity to be able to begin at the beginning, unfettered by custom, convention, and competing opinions in his consideration of what is true and what is not. Absent this possibility, Descartes proceeds with what would be second best, a kind of "second sailing."[15] Descartes thus attempts to replicate what it would mean to begin at the beginning, despite never actually being able to do so. The notion of methodic doubt, and method generally, points exactly in that direction.

Now, as Part II of the *Discourse* reaches its end, Descartes articulates the implications of the titular method of the book: "What contented me most about this method was that by means of it I was assured of using my reason in everything, *if not perfectly*, then at least as well as was within my power"

(II.13, emphasis added). The substance of Descartes's discussion of method—more precisely, of four precepts of the method—that takes up the second half of Part II cascades directly into Part III, for the latter begins quickly and quirkily with the words, "And finally," as if it were a direct and uninterrupted continuation of the prior part's thought. Descartes, then, immediately likens what he will be proposing—a provisional morality—to one's need "to make provision" (*faire provision*) for temporary lodging during the time between tearing down one's house and finally rebuilding it: "Thus, so that I would not remain irresolute in my actions, while reason obliged me to be so in my judgments, and that I did not cease to live from that time on as happily as I could, I formed for myself a provisional morality (*une morale par provision*), which consisted of only three or four maxims, which I would like to present to you" (III.1). Without proceeding any further, however, attentive readers of the *Discourse* will have noticed that what Descartes delineates in these lines is at variance with the "Synopsis," which informs us that we will find in Part III "some rules of morality," and that these will be drawn from the method. Thus, instead of a morality, we encounter a *provisional* morality; instead of rules, we find *maxims*;[16] rather than *some* rules, we find *three or four* maxims that seem to be exhaustive; and what might be thought of as the fourth maxim is, in fact, no maxim at all, but rather articulates Descartes's choice of life, the best way of life, philosophy or science. Presumably, all of these discrepancies matter and are intended. But what matters most for us here is that in order to free himself from the confines of custom and convention, Descartes finds it necessary to find a place for himself within those confines. It is only by doing so that he can situate himself to be best able to distinguish true from false judgments.

Let us begin, naturally enough, with the first maxim: "To obey the laws and customs of my country, constantly adhering to the religion in which God has given me the grace to be instructed since my childhood, and governing myself in every other thing by following the most moderate opinions, and most removed from excess, which were commonly accepted in practice by the most sensible of those with whom I would have to live" (III.2). Descartes is seeking to know how to be resolute in his actions during the time in which he is irresolute in his judgments, as he calls his opinions before the tribunal of reason. However, as the tail of this maxim implies, action rests upon opinion or belief. So, despite his initial statement, it is not so much here that he is seeking a series of actions to pursue as that he is seeking a series of opinions to guide him. Even the laws, customs, and religion above are practices that are structured and animated by a variety of opinions. And one inference we can draw is that, for Descartes, since the meanings and requirements of laws, customs, and religion are not self-interpreting or self-evident—and it is somewhat bold to set forth obeying one's religion as part of a *provisional* morality—he would have to rely on the opinions of the most sensible interpreters for an interpretation of them.

So, we need to know who the most sensible interpreters are. However we are to identify them, Descartes indicates that they are relative to place: "And although there may perhaps be as many who are sensible among the Persians or the Chinese as among ourselves, it seemed to me that the most useful thing was to regulate myself in accord with those with whom I would have to live" (III.2). The most sensible of the Persians and Chinese, of course, would have a different religion they would follow, to say nothing of different laws and customs, than the people with whom Descartes was living. The reason, according to Descartes, for not following all of these, though, has everything to do with location, and nothing to do with whether they are true, just, noble, or laudatory in some other sense. But in a certain way, Descartes's interest is less in identifying those who are the most sensible and more in identifying how to ascertain what their opinions are, for what they *say* their opinions are is not something that can be trusted. One reason why is that "due to the corruption of our morals there are few people who would want to say all that they believe" (III.2). Descartes does not state, of course, what that corruption is, but certainly it has to be connected to issues of freedom of thought and freedom of speech, and, just as certainly, it has to be connected to the circumstances of his own time, toward which Descartes gestures in Part VI, where, when tacitly commenting on a certain understanding of Galileo's, he uses the exact same phraseology: "I do not want to say that I agreed with it" (VI.1).[17] We should also take note of Descartes's choice of pronoun here: the corruption in question is of *our* morals, and this presumably refers back to "ourselves," those to whom he is comparing the non-European, non-Christian Persians and Chinese.[18] But another reason Descartes offers as to why we cannot trust the most sensible to state their opinions accurately is that, although people believe many things, many do not *know* what they believe. Thus, in the light of these two reasons, we are led to pursue the opinions of the most sensible indirectly, through their actions.

Nonetheless, Descartes comes to recognize that those he has denominated the most sensible might hold differing opinions on the same subject, in which case he will have to choose among their divergent views. Conventional opinion—for that is what the opinions of the most sensible represent—is not always uniform. This recognition leads Descartes to focus entirely on the character of the opinions that he might follow, and not at all on those who might hold them: "And among several opinions equally accepted, I chose only the most moderate, partly because these are always the most convenient in practice, and probably the best, all excess customarily being bad" (III.2). Descartes, as if by philosophical prestidigitation, presents himself as being able to ascertain the most moderate opinions, as well as those opinions suffering from excess, presumably on the basis of some feature of those opinions themselves. Indeed, in claiming to be able to choose the most moderate opinions to follow, Descartes seems to be reaffirming what he intimated earlier about "governing myself" with respect to the opinions of the most sensible.

The compass of Descartes's presentation of his first maxim steadily points him in the direction of what is commonly accepted and practiced, especially moderation. However, as if a magnetic storm has swept through, rendering that compass unreliable, Descartes stealthily, in the name of moderation, begins to navigate the reader in a completely different direction, with new sights to behold. He tells us, in virtue of upholding moderate opinions, that he "placed among the excesses all the promises by which one curtails something of one's freedom." An immediate qualification follows, though: laws, such as those involving contracts, commerce, and vows, which oblige one to persist in an endeavor, are acceptable to him, if only because they are a remedy for "the inconstancy of feeble minds" (III.2).[19] This recasts Descartes's opening statement about obeying the laws and customs of his country; also, in that very statement, he spoke of "constantly" adhering to the religion with which God had graced him, which here must be linked to the "inconstancy" of the weak minded, which leads them to require laws.

But Descartes now moves ahead and daringly asserts he cannot countenance the abridgement of his freedom: "Because I did not see anything in the world that always remained in the same state, and because, for my part, I promised myself to perfect my judgments more and more, and not to render them worse, I would have thought I was committing a great offense against good sense if, because I had approved of something once, I was obliged to take it to be good still later, when it would have perhaps ceased to be so, or when I would have ceased to esteem it to be such" (III.2). Just as, from a certain perspective, everything is in motion, so too must Descartes's judgments be allowed to be in motion, even if one or more of them temporarily has come to rest upon a certain understanding: Descartes must be free to vary his judgments for a varied world, as he sees fit. Of course, this looks something like the undertaking he announced in Part II of the *Discourse*, of calling all of his opinions into doubt, which would require the freedom to reject some, but a freedom to accept others when they are raised up by means of applying a proper standard of reason to them. However, a striking difference here is something akin to a moral lexicon that Descartes brings to bear on his activity: he has "promised" himself to improve his understanding; he would be committing a great "offense" or "fault" if he did not change his mind when warranted; and there is an approach he is "obliged" to take—all apparently implicated by the activity of reason. But all of these commitments rest on a condition of freedom. Noticing all of this brings into sharp relief a tension that arises with Descartes's appeal to promises: on the one hand, he evinces a concern about promises that curtail his freedom; while the other hand, he expresses a promise to perfect his judgments more and more, which he willingly takes on, and which, rather than curtailing his freedom, appears to enable it.

The passage we have been examining, which brings the paragraph containing Descartes's discussion of the first maxim to an end, can be seen to come to light not as part of an elaboration of that maxim itself, but what might be understood as an "effectual retraction" of it.[20] However, rather than seeing Descartes's commitment to being free—that is, being able to call his own opinions as well as those of others into question—as a retraction of the maxim, it is rather that, given the shadows cast by the world in which he lives, setting down the maxim, and at least appearing to be guided by it, serves to bring to light those conditions under which he is able continually to shape his judgments and opinions in accordance with reason. This tension itself thus reveals that Descartes's freedom to set custom, convention, and common opinion aside in the pursuit of truth requires that he be cloaked by some of these customs, conventions, and common opinions themselves.

Descartes's discussion of the second maxim pivots away from questions having to do with his opinions and true and false judgments—that is, with questions about thought—and toward questions directly having to do with action: "My second maxim was to be the most firm and the most resolute in my actions that I could be, and *to follow* no less constantly the most doubtful opinions, when I had once determined to do so, than if they had been very assured" (III.3, emphasis added). He follows that remark by likening himself to travelers lost in the middle of a forest, who should not wander around, going this way or that, but should follow a straight path to wherever it leads, even if that path were decided by chance alone. Although they might not end up where they desire, they "probably" will be better off than remaining in the middle of the forest. Descartes, then, precisely in the middle of the paragraph, begins the very next sentence "And thus," as if he were either continuing the same train of thought or drawing a conclusion from it, only to affirm that since the actions of life often do not permit of any delay, if we cannot ascertain the truest opinion to guide our action, we should rest our action upon the most probable; and if we find several opinions equally probable, we should decide on one, and not regard it as "doubtful" but as "true and certain" (III.3), irrespective of the fact that it would be simply probable. As pertains to human action, the ground for pursuing what is useful in life is generally found in the realm of probability and not certainty.

Despite the appearance of continuity, though, the second half of the paragraph is in marked tension with the first. For whereas the first half directs our attention to the most *doubtful* opinions and following them, once we had determined to do so, no differently than had they been *very assured*—two poles *in extremis*, with no middle ground of any kind—the second half directs our attention to probability, some form of middle ground. Moreover, whereas Descartes's statement of the maxim proper speaks of him alone, as he continues, and probability in action instead of truth in judgment comes to the forefront, he repeatedly invokes plural pronouns. Nevertheless, at the very end of the paragraph, he turns back again to himself and (just as was the case with first maxim) he concludes with an appeal to his own freedom—and his own freedom alone, for acting based upon probability "was able from that time onwards *to deliver me* from all the repentance and remorse that customarily agitate the consciences of those feeble and unsteady minds, who allow themselves to proceed inconstantly in practicing as good the things that they later judge to be bad" (III.3, emphasis added). One can see here that the language of Christendom permeates the beginning and end of Descartes's articulation of the second maxim. There is, just as is the case with the first maxim, a subtle understanding of bondage at work, one Descartes tightly ties to Christianity. Yet in both instances, Descartes finds it necessary to embrace both maxims in order to be free, both with regard to his judgments and with regard to his actions.

Let us now turn to the third maxim, which functions differently from the first two insofar as it directly concerns mastery and the complicated relationship between thought, action, and nature that bears on the overall intention of the *Discourse* as a whole: "My third maxim was always to try to conquer myself rather than fortune, and to change my desires rather than the order of the world; and generally to accustom myself to believe that there is nothing that is entirely within our power except our thoughts, so that after we have done our best concerning the things that are external to us, all that which is lacking for us to succeed is, with respect to ourselves, absolutely impossible" (III.4).[21] Here, Descartes gives primacy to affecting or transforming himself rather than fortune or the world—and this is no doubt in keeping with what had been the title of Descartes's book up until about one year before it appeared: *The Project of a Universal Science that is Capable of Raising Our Nature to its Highest Degree of Perfection*.[22] Yet such an orientation might be seen to conflict with the overall humanitarian or benevolent thrust of the *Discourse* that Descartes announces in Part VI. There he commits himself to a "practical" philosophy, useful for this life, in opposition to "the speculative philosophy that is taught in the schools" (VI.2). Part and parcel of such a philosophy is knowing and learning how to employ the forces of nature for the amelioration of the human situation, thereby affording us the possibility of "making ourselves like masters and possessors of nature" (VI.2), in pursuit of the human goods. Descartes's third maxim, however, occludes from view the reformatory character of such an endeavor.

Nevertheless, this maxim is tantalizingly ambiguous. That Descartes here privileges changing himself over changing the world does not require that he discount the importance of the latter: in fact, given the way the whole of the argument in the paragraph at hand unfolds, it rather appears to be the case that "conquering" himself provides a condition for the latter.[23] Moreover, if it is only his own thoughts that are *entirely* within his power, this does not entail by any means that other things are not also within his power. They simply depend on still other factors: for example, our coming to know something about them; coming to know how to act upon them in the world; and various institutional arrangements, *et cetera*. As we reach the tail of the maxim, Descartes's emphasis is on doing *our best* to change the external world, and if we have done that yet still fail, we can say that what we set out to do is impossible; but it is not obvious what counts as doing our best and how to know when we have done so. Indeed, Descartes cleverly leaves hanging whether that failure lies in matters exterior to us, or whether the responsibility lies with us.[24]

One might be tempted to say as a shorthand that the third maxim presents itself as a kind of muted Stoicism. Nevertheless, as Descartes weaves his way through the rest of his discussion of the maxim, it is ancient Stoicism in all of its vigor that marches prominently into view: "If we consider all the goods outside of us as equally removed from our power . . . and making a virtue of necessity, as one says, we will no more desire to be healthy when sick, or to be free when in prison, than we now do to have bodies of a material as little corruptible as diamonds, or wings to fly like birds" (III.4). Such a hyperbolic approach to life cast, it must be added, completely in the conditional mood—an element of which sees us as having no desire to be angels (hence the reference to incorruptible bodies and wings to fly)—would require long meditative practice, wherein lies "the secret of those philosophers who were able in former times to escape the empire of fortune and . . . to vie with their gods for felicity" (III.4). And such practice, which no doubt draws from a certain understanding of freedom, will be able to persuade them that "nothing was within their power except their thoughts" (III.4). We have gone, then, from concerns about conquering fortune to dedicating oneself to escaping it altogether; and from the position that our own thoughts alone are *entirely* in our power, to *nothing but* our thoughts are ever in our power. Whatever we might say of the statement of the maxim with which Descartes begins, we can surely say that the philosopher with whom Descartes ends has no interest at all in the reformation of the world or transformation of nature; but Descartes's artful presentation encourages the reader to see the entirety of the paragraph as being of a piece. This is no doubt done by Descartes with the intent to make it appear that his only concern is with himself, and he is content with withdrawing from the world. But having seemingly resigned himself to the world as it is, Descartes then turns to the so-called fourth maxim.

Descartes introduces his subject not as a maxim but as "the conclusion of this morality." He recounts, here in the middle paragraph of the *Discourse*, how he weighed the different occupations that men have in this life so as to be able to choose the best one. He thought he could not do any better than to carry on in what he was already doing, which was "employing all my life in cultivating my reason, and in advancing as much as I could in the knowledge of the truth, following a method I had prescribed for myself," a life that had already brought Descartes "extreme contentment" (III.5). Discovering new truths every day, he tells us, thoroughly filled his mind, and nothing else concerned him. But after this declaration, Descartes returns to the maxims of the prior pages: "The three preceding maxims were founded only on the plan to continue to instruct myself; for since God has given each of us some light to discern the true from the false, I would not have believed that I should be content for a single moment with the opinions of others, if I had not proposed to employ my own judgment to examine them when the time came" (III.5). The three provisional moral maxims, then, have been formulated by Descartes to work in the service of the best of human occupations for this life, philosophy or science.[25] Having delineated this understanding, Descartes goes on to assert that he would not have been capable of being content without following the occupation or path he had chosen, for only by that means would he be able to acquire all the knowledge of which he was capable, as well as acquiring "all of the true goods that would ever be in my power" (III.5). This seems a far cry from the image of the Stoic philosopher—both muted and otherwise—that Descartes put on display during his discussion of the third maxim, and it appears more in keeping with the humanitarian, scientific-technological enterprise that Descartes articulates in Part VI.

However that may be, as we have seen, the first three parts of the *Discourse* again and again come face to face with custom, convention, and common opinion. Above all else, these encounters unite the first half of the work. Indeed, toward the end of Part III, Descartes indicates that his thoughts concerning a provisional morality were more or less coeval with the thought with which he began Part II,[26] a linkage connected to the fact that these two parts are framed by war.[27] Further connected here is that, at the head of his presentation of the third maxim, rather than using the language of self-examination or mastery, Descartes writes of *conquering* himself: there is a war afoot in which Descartes is a philosophical combatant.[28] Nevertheless, Part IV attempts to lay the ground for a new beginning, and the thematic landscape of custom and convention seems to be abandoned as Descartes waxes metaphysical. Yet, the final line of Part III provides a conduit to that new beginning and establishes a thematic ligature binding Parts III and IV together.

Descartes tells us that nine years of travel followed that crucial time spent during winter in the *poêle*. But after that, he withdraws and settles in "a country where the long duration of war has established such order that the armies maintained here seem only to enable one to enjoy the fruits of peace with so much more security, and where amidst the crowd of a great, very active people, and more attentive to their own affairs than curious of those of others, while lacking none of the conveniences of the most populous cities, I have been able to live as solitarily and withdrawn as in the most remote deserts" (III.7).[29] While it is worth noting that the unnamed country is the

United Provinces (the Netherlands), it is more important to identify what Descartes actually is saying and doing at the end of the passage. Here, no translation can capture the artfulness of Descartes's French, which reads, "*les déserts les plus écartés.*" This turns out to be a play on words akin to an anagram, as depicting it in this form reveals—*les DÉSerts les plus éCARTÉS.*[30] Descartes was fond of anagrams,[31] and in the seventeenth century it was usual for his name to be spelled, both by himself and others, as Des Cartes. So here we have him, representing himself as if he were in the most remote of deserts, a wilderness, and thus apart from all common opinions and conventions, conceiving the possibility of delivering or liberating himself from the obstacles to understanding, but also Christian Europe from the political-theological stranglehold from which it was literally dying.

After Descartes learned of the narrow straits in which Galileo was floundering, he wrote, in 1634, to his friend Mersenne to make known his desire "to live in repose and to continue the life I have begun, taking for my motto, *He lived well who hid well*"—a line silently borrowed from Ovid's *Tristia* (III.4.25).[32] Two things are thus intertwined at the very close of Part III: Descartes inscribing his name, that is, himself, in the *Discourse*, hiding himself in it; simultaneously, through the very same words, he implies that he is as hidden while in a vast, populated area as if he were in a vast desert. All of this raises questions about the "I" that Descartes will shortly showcase.

The opening of Part IV of the *Discourse* is as well known as any piece of philosophical literature. After announcing his metaphysical intentions, Descartes is soon led, through a process of doubt, to what he terms the first principle of the philosophy that he was seeking—"I think, therefore I am" (IV.1).[33] And while he was able to "feign" that he had no body, he could not help but affirm that he was "a substance the entire nature or essence of which was to think." But what to call such a substance? While doubting the veracity of the senses along with other faculties, Descartes spoke of the activity of the mind; however, that is not what he identifies as being this substance. Instead, choosing a term at home among a litany of Christian doctrines, he declares himself to be a soul: "So that this me, that is to say, the soul by which I am what I am, is entirely distinct from the body" (IV.2).[34] Descartes offers no reason or argument to justify identifying himself with a soul—he merely asserts it. Despite this, the "Synopsis" tells us (and Descartes intimates) that one will find in Part IV a demonstration or proof for the existence of the soul. Given that Descartes is in the midst of calling all of his prior opinions into question, such an identification seems rather unwarranted.

Let us not, though, be troubled by it here. To do so might lead us astray from what is perhaps the most remarkable statement in all of the *Discourse.* For in identifying himself with "the soul *by which I am what I am*" (emphasis added), Descartes is in fact adopting a passage, albeit in French, from the Vulgate (Latin) edition of The Book of Exodus. We read there of the Lord saying to Moses that he has seen the affliction of "my people in Egypt.... And knowing of their sorrow I have come to deliver them out of the hands of the Egyptians and to bring them out of that land... into a land that floweth with milk and honey" (3:7–8). The Lord tells Moses that he is sending him to the Pharaoh so that Moses can bring His people out of Egypt, freeing them, but Moses is concerned lest the people ask the name of He who sent Moses to them, for he would not know what to say in response. And God said to Moses that this is what you should say of me: "I am who I am" (3:14).[35] The Vulgate's "*ego sum qui sum*" maps on precisely to Descartes's "*je suis ce que je suis*"—indeed, so precisely that we would be hard-pressed to maintain that it was not part of Descartes's intention to formulate the matter as he does. If we are correct in drawing this connection, then we can see the anagrammatic end of Part III and this remark from Part IV as binding the two halves of the work together, forming a thematic ligature of liberation. Descartes, here, in one middle of the *Discourse*, appears as both Moses and God: Moses, apparently insofar as he withdraws to a desert in order to deliver himself and his people from the chains of custom, convention, and common opinion; and God, apparently insofar as he is representative of beginning at the beginning and of being the creator of a world—here, though, a new world. The six parts of Descartes's *Discourse*, then, can be seen as emblematic of the creation in Genesis over six days, and, indeed, Descartes parallels *that* creation in Part V, beginning with light and ending with man.[36] But in likening himself to God, is Descartes pointing to something still more?

In pursuit of that question, let us notice that Part IV appears to be set down so as to establish the foundations on the basis of which Descartes's physics can be adumbrated and justified. We have been prepared to see, then, a careful analysis taking us from the proofs for the existence of God and the soul, on through other principles, until we are linked up with and arrive at Descartes's physics. However, such an expectation is dashed at the head of Part V, as Descartes reveals that the matters connecting his metaphysics and his physics are so "controversial among the learned" that it would be better for him to refrain from presenting them. But after indicating his allegiance to the principle of philosophy he discovered in the prior part, and to accepting no judgment as true unless it satisfied certain criteria, Descartes turned to the subject of the laws of nature: "I dare to say that not only have I found the means of satisfying myself in a short time concerning all the principal difficulties that are customarily treated in philosophy, but also that I have noticed certain laws that God has so established in nature, and of which he has imprinted such notions upon our souls, that after having made enough reflection on them, we could not doubt that they are exactly observed in all that which is or all that which takes place in the world" (V.1).

Descartes then informs the reader why he has refrained from publishing *The World*,[37] the book on which he was working, but that at this point in the *Discourse* he will submit some of it to the light of day, while of necessity keeping other parts in the shadows. In particular, he will submit some aspects of his teaching on the subject of light, as well as on the sun and stars, because they emit light, and all bodies on earth, because they reflect light, and man himself, because he is a spectator of it. However, since Descartes wants to avoid various disputes, he chooses to leave the current world aside, and to present an imaginary account—which in the corresponding analysis in *The World* he calls a "fable"—of what would happen if God were to create a new one.[38] Of course, once "created," this *fabulous* world looks and acts exactly like the one Descartes understands himself to inhabit; but our interest is in the creation itself and, in particular, the role of the laws of nature in that creation.

Imagine, then, what would happen in a "new world," Descartes bids us to consider, "if God now created somewhere, in imaginary spaces, enough matter to compose it, and that he agitated in different ways and without order the diverse parts of this matter, so that he composed there a chaos as confused as the poets could feign, and afterwards did nothing else but lend his ordinary assistance to nature, and let it act according to the laws that he has established" (V.2).[39] Now, the opening paragraphs of Part V are consistent in this regard: what God creates is chaos, and only after the creation, Descartes affirms, does God establish the laws of nature, which render that chaos orderly—that is, it is not the creation that brings forth order, but instead it is the "establishment" of the laws of nature that does so. Furthermore, Descartes asserts that however many worlds God may create, "there could not be any of them in which these laws failed to be observed" (IV.2), and thus all of the possible worlds God might create would have to be fundamentally the same.

But as Descartes alludes to the laws of nature in the *Discourse*, he continually refers to how, in the book he never published, he describes with precision the way in which these various laws order the chaos that is before them. When one combines this with Descartes's claim that God *imprinted* the laws of nature upon our souls—a claim in tension with the fact that, prior to Descartes, a discussion of scientific laws of nature was extremely rare,[40] as well as the fact that very few individuals have even the slightest conception of them—a certain portrait of the laws of nature begins to emerge. The *laws* of nature are, for Descartes, in some large measure, products of the mind, instruments, by which the "chaos" of the material realm can be systematized and understood. On this analysis, chaos can be understood to be less a feature of the world itself, and more a feature of the world not understood. Laws of nature, especially once they can be mathematically expressed,[41] impose a certain order upon the world whereby it can be both understood and, ultimately, manipulated for the amelioration of the human situation. Perhaps an early clue to this understanding occurs in Part II, in Descartes's discussion of the third precept of the method, where he somewhat opaquely asserts that in starting with the simplest of objects and proceeding little by little to the more complex ones, he will be "supposing an order among those [things] that do not naturally precede one another" (II.9). The laws of nature, then, might be said to be less a matter of discovery, and more a matter of being devised. The laws of nature, to put it otherwise, require a lawgiver, and Descartes recognizes himself to be rather uniquely suited for such a role.

These comments on the laws of nature lead us naturally enough to Descartes's Baconian claim at the beginning of Part VI of the *Discourse*: equipped with an understanding of the elements, coupled to an understanding of the crafts of various artisans, "we might be able, in some fashion, to employ them in all the ways that are appropriate to them, and thus to render ourselves as masters and possessors of nature" (VI.2).[42] If we are able to order nature, then we seem to be in a good position to give orders to it and thereby reorder it. The "Synopsis" of the *Discourse* informs us that it is in the sixth part that the author of the work makes known his reason for writing the book. That reason is a benevolent or humanitarian one: to improve the human lot in life. And what should be clear is that Descartes's appeal to the "general good of all men" is aimed at being as universal as the parallel appeal of Christianity. In staking out such a claim, Descartes intends to create a model of a good life that can serve as a competitor to Christianity, but one rooted in, and aimed at, "this life,"[43] one in which one's body figures prominently. As "masters and possessors of nature," we might be able to "invent an infinity of artifices," the most important ones of which would work toward what Descartes prescribes as "the first good and the foundation of all the other goods of this life," namely, "the preservation of health" (VI.2).

Yet, we must ask whether Descartes's proposal of a scientific-technological enterprise, which carries with it the promise of transforming nature for the improvement of the human situation, is to be taken straightforwardly, or is it rather hyperbolic, a rhetorical device aimed in part at taming Christianity, or at least some of its excesses, by offering a this-worldly alternative, one still full of hope and aimed at mitigating fear? Is the *Discourse* a reflection of a man who, while he does not want to fall "into the extravagances of the paladins of our romances" (I.8), nevertheless is offering a different extravagance, one that takes the form of a counterweight to Christianity? Early in the *Discourse*, Descartes informs his readers that he "proposes this writing only as a history or, if you prefer, only as a fable" (I.5). And while fables "awaken the mind" (I.7), they also "make someone imagine many occurrences as possible, which are not" (I.8). Upon reflection, we must decide whether the offer of the possibility of becoming as masters and possessors of nature is part of what might make the *Discourse* a fable: is this an idealization akin to the idealizations that populate the opening paragraph of Part II? How much, in other words, might some semblance of mastery over ourselves, especially with respect to our judgments of true and false, carry over to mastery over nature? However we are to settle these questions, which are at the heart of understanding the *Discourse*, in doing so we must give proper weight to Descartes's so-called fourth maxim, the discussion of which occupies the middle paragraph of the book, in which Descartes affirms that a life spent cultivating his reason in pursuit of the truth is the very best of occupations.[44]

Chapter 5

Montesquieu, Commerce, and Science

Diana J. Schaub

Montesquieu is well known as an advocate for commerce, stating in *The Spirit of the Laws* (1748) that "commerce cures destructive prejudices."[1] Less known—indeed, almost unknown—is that Montesquieu made the identical claim about science. In a short work entitled "Discourse on the motives that ought to encourage us to the sciences," delivered before the Academy of Bordeaux in 1725, Montesquieu asserts that "the sciences are very useful in that they cure peoples of destructive prejudices."[2] The striking similarity gives rise to a host of questions. Since almost a quarter of a century separates the two claims, did Montesquieu abandon the first claim about science in favor of the second about commerce, or did he remain committed to both claims? Are the claims fully distinct, or is there perhaps some essential connection between science and commerce, such that the second claim is an extension of the first? Finally, is the thing being cured—the "destructive prejudices"—the same or different in the two cases?[3]

Before examining each claim in its textual setting, it might be worth noting that curing (or healing) prejudice is absolutely central to Montesquieu's enterprise. Perhaps that is not surprising given that the critique of prejudice figured in the Enlightenment from the start, with both Bacon and Descartes inveighing against prejudice as the primary obstacle to the ideal of detached scientific judgment. Thus, in Bacon's *Novum Organum*, "prejudice" is the comprehensive term for the "errors, false notions, superstitions, and delusions" that afflict the human mind.[4] For Descartes too, "precipitancy and prejudgment are held up as the great offenses against philosophizing."[5]

Montesquieu, however, gives a somewhat different definition, scope, and application to the word "prejudice."[6] Consider the multiple references to "prejudice" just in the two-page Preface to *The Spirit of the Laws*.[7] With bold assurance, he declares: "I did not draw my principles from my prejudices but from the nature of things."[8] Yet, there is an element of humility in this pronouncement, since Montesquieu does not claim to be prejudice-free; he has *his* prejudices. His claim is the more limited one that, over the course of twenty years' labor, he managed to extract principles from "the nature of things" without undue interference or contamination from his prejudices. Thus, his prejudices do not prejudice his principles.

If this is true, then Montesquieu differs from those philosophers whom he cites in the chapter entitled "On Legislators," which, until the late addition of the four books on Roman and French history, was intended to be the very last chapter of this massive work.[9] Here, Montesquieu offers startlingly dismissive, one-sentence summaries of the personal motivations of Aristotle, Plato, Machiavelli, More, and Harrington, leading him to conclude that "the laws always meet the passions and prejudices of the legislator. Sometimes they pass through and are colored; sometimes they remain there and are incorporated."[10] Given that the examples are of political philosophers rather than either ordinary legislators or even great founders, one wonders how this prejudicial tint would not apply to Montesquieu himself. Perhaps one way to square the claim of disinterestedness made in the Preface with the final claim about the partiality of all legislators is to conclude that Montesquieu is not a legislator, not even of the philosophic sort. After all, Montesquieu's subject is not laws but the spirit of laws—the necessary propaedeutic to legislation but not itself legislation. Part of understanding the spirit of laws includes awareness of the passions and prejudices of the heretofore guiding spirits of antiquity and modernity. Another possibility is that, until Montesquieu, no philosophic legislator possessed the spirit that ought to belong to the legislator: namely, the spirit of moderation, which effectively counterbalances or neutralizes partiality.[11] In an extraordinary passage, buried in 28.41 on "The Ebb and Flow of Ecclesiastical Jurisdiction and Lay Jurisdiction," Montesquieu explains the immoderation of great men:

> By a misfortune attached to the human condition, great men who are moderate are rare; and, as it is always easier to follow one's strength than to check it, perhaps, in the class of superior people, it is easier to find extremely virtuous individuals than extremely sage ones.

> The soul takes such delight in dominating other souls; even those who love the good love themselves so strongly that no one is so unfortunate as to distrust his good intentions; and, in truth, our actions depend on so many things that it is a thousand times easier to do good than to do it well.[12]

Montesquieu, who is perhaps one of these rarest of self-restraining souls, hopes to improve the odds of "doing good" well.

Principles drawn from "the nature of things" provide the key.[13] "The nature of things" is a phrase greatly favored by Montesquieu.[14] It occurs prominently in Bacon too; witness the précis that Bacon places at the head of his Preface to *The Great Instauration*: "*That the state of knowledge is not prosperous nor greatly advancing; and that a way must be opened for the human understanding entirely different from any hitherto known, and other helps provided, in order that the mind may exercise over* ***the nature of things*** *the authority which properly belongs to it*" (italics in original, emphasis added).[15] Bacon promises an undeniably expansive exercise of authority, including "all power of operation."[16] By contrast, Montesquieu, in keeping with his greater moderation (and his focus on political and constitutional matters), suggests that the greater one's knowledge of "the nature of things," the greater should be one's caution in exercising authority. Since "everything is extremely linked,"[17] it is exceedingly difficult to grasp the articulated whole and thus to anticipate all the ramifications of intentional change. In the Preface, Montesquieu notes that a very conservative conclusion "will naturally be drawn": namely, that "changes can be proposed only by those who are born fortunate enough to fathom by a stroke of genius [*un coup de génie*] the whole of a state's constitution."[18] If that is the prerequisite for allowable change, then such change will be rare indeed. Montesquieu's twenty years of frustrating study—"a thousand times I cast to the winds the pages I had written; every day I felt my paternal hands drop"[19]—is enough to prove that his vast understanding did not arise in this wondrous way.[20]

While Montesquieu is aware that conservatives will jump to a conservative conclusion (and embrace him as one of their own), that does not seem to be Montesquieu's own conclusion. He is far from a wholesale dissenter from the project of enlightenment. As he goes on to say: "It is not a matter of indifference that the people be enlightened."[21] Thus, prejudice is his target. Both rulers and ruled are implicated, since "the prejudices of magistrates began," according to Montesquieu, "as the prejudices of the nation."[22] Still, reform seems to be stymied, since Montesquieu contends that those who are enlightened are as wary of change as is "each nation" (which naturally upholds its established way of life and may do so more intelligently after learning from Montesquieu "the reasons for its maxims"[23]). Here is his description of the conservatism of the enlightened: "One feels the old abuses and sees their correction, but one also sees the abuses of the correction itself. One lets an ill remain if one fears something worse; one lets a good remain if one is in doubt about a better."[24]

Despite the fears and doubts that counsel against corrective action, Montesquieu does not abandon his ambition to work an instructive change in mankind. In a stylistically brilliant series of three conditional hypotheticals, he sketches the elements of his plan.[25] The series culminates in a statement elucidating Montesquieu's complex approach to "prejudices": "I would consider myself the happiest of mortals if I could make it so that men were able to cure themselves of their prejudices."[26] Montesquieu proposes a unique form of cure: human beings will cure themselves, and yet it will be Montesquieu who makes them do it. How could this be? Much depends on the definition of prejudice, and Montesquieu offers his: "Here I call prejudices not what makes one unaware of certain things but what makes one unaware of oneself."[27] This seems a narrower definition than, say, Bacon's, since it is limited to what impedes self-awareness rather than encompassing all "the idols of the mind" that interfere with the comprehensive knowledge and mastery of nature. On the other hand, to "know thyself" in full may entail an all-inclusive quest.[28]

In any case, Montesquieu is confident that man is "capable of knowing his own nature when it is shown to him."[29] Fascinatingly, Montesquieu hints that our self-knowledge has been stolen or concealed from us as a result of a peculiarity of our nature. Man is "that flexible being"[30] whose social pliancy is such that he can be led to forget himself. Human beings regularly exchange their natural flexibility for the artificial fixity of particular conventions. Despite Montesquieu's assurance that he does not have "a censorious spirit,"[31] a reading of *The Spirit of the Laws* reveals that many of these characterological metamorphoses are profoundly misshapen—afflicting and distorting important aspects of human nature. Montesquieu's aim is restorative. The results of a return to self-awareness and self-possession are described in the first two conditionals: when rulers learn to craft better laws, the ruled will discover "new pleasure in obeying" and "new reasons for loving" their laws.[32] In other words, each nation will not simply rest content with understanding "the reasons for its [existing] maxims," it will instead acquire "new reasons" that better accord with human nature as Montesquieu intends to show it.

His "showing" takes the form of a comparative treatment of the many instantiations of self-concealment. By canvasing "the histories of all nations," Montesquieu creates an incredibly densely textured analysis. One is tempted to say that he is at home wandering the infinite, which for him means "the infinite diversity of laws and mores" (mirrored in "the infinite number of things in this book").[33] His method seems the counter-pole to the stripped-down "state of nature" approach of the other early moderns. In that same Preface (which cannot be read too closely), Montesquieu, without naming names, disparages the "salient traits" so characteristic of "present-day works."[34] Montesquieu does not say that these aspects, so prominent in the work of others, will be absent from his, only that these features will lose their simplicity or singularity by being placed in fuller perspective.

Given his manner of proceeding, it is not surprising that Montesquieu is usually not regarded as part of the "state of nature" crowd. Nonetheless, Montesquieu does make rare, but highly significant references to a "state of nature." The first time occurs very early, in book 1, chapter 2, as he sets forth "the laws of nature," distinguishing his views, ostentatiously, from those of Hobbes.[35] The second reference comes in the last book of part I, where, quite out of the blue sky, he links "the true spirit of equality" to an orthodox account of the state of nature: "In the state of nature, men are born in equality, but they cannot remain so. Society makes them lose their equality, and they become equal again only through the laws."[36] Finally, hundreds of pages later, near the end of the journey, Montesquieu three times invokes the state of nature. Those readers who have not fallen exhausted by the wayside learn how the German peoples contractually exited the state of nature. We are given a concrete instance—indispensable, Montesquieu says, for understanding "our political right"—of how aggrieved private parties reached "reciprocal agreement." These "*compositions*" (or "settlements") then became the model for sagacious rulers to draft public and obligatory versions of the private settlements.[37] Montesquieu concludes: "By establishing these laws, the German peoples came out of that state of nature in which it seems they still were at the time of Tacitus."[38] It would be hard to underestimate the importance of this description, since at the close of the famous chapter on the English constitution (11.6), Montesquieu rather cryptically declared that "the English have taken their idea of political government from the Germans. This fine system was found in the forests."[39] Only when we reach book 30 does Montesquieu locate the source of the regime of liberty in an original contract, or rather in an existing practice of devising "satisfactions" that was subsequently rationalized and generalized.

Montesquieu's method—widely separating linked passages and blunting salient features—requires readers to act as compositors. The labor becomes an education in moderated modernity. We are invited into the exercise of Montesquieu's wellness program for political life. My hunch is that Montesquieu was concerned about the disruptive potential of the Lockean equation (equality + rights + consent = revolution). Accordingly, he downplayed its constitutive elements without precisely repudiating any of them.[40] With respect to other fellow moderns, he was a bit more forthright, directly voicing concerns about their immoderation. This is especially the case with Machiavelli, Bayle, and Spinoza.[41] The one exception seems to be Descartes, of whom he always spoke well.[42] This anomalous praise brings us around finally to that intriguing remark about the curative effects of science on "destructive prejudices."

Montesquieu begins and ends the discourse on science with the name of Descartes. Here is the key passage: "If a Descartes had come to Mexico or Peru one hundred years before Cortez and Pizarro, and if he had taught these peoples that men, composed as they are, are not able to be immortal; that the springs of their machine, as those of all machines, wear out; that the effects of nature are only a consequence of the laws and communications of movement, then Cortez, with a handful of men, would never have destroyed the empire of Mexico, nor Pizarro that of Peru."[43] Interestingly, Montesquieu does not attribute the rapidity of the European conquest to any technological superiority resulting from the scientific revolution. Instead, he presents the European conquest as an American collapse. The Aztec and Incan empires crumbled, despite real defensive advantages, because of their "ignorance of a principle of philosophy."[44] When startled by anything "new," like the aspect and appurtenances of the strange visitors, the natives were left paralyzed by their belief in "power invisible"[45]—a belief that had been instilled by their despotic rulers: "To make themselves revered as gods, the Princes had made their peoples as stupid as animals, and they perished by that same superstition that they had authorized for their advantage."[46] Political power—and crucially, the moral wherewithal for collective defense—is tied to theological assumptions. This judgment is echoed in *The Spirit of the Laws*, where Montesquieu says that "the prejudices of superstition are greater than all other prejudices."[47]

Cartesian science is valuable because it gives the lie to the godlike pretensions of rulers.[48] One is reminded of Jefferson's deathbed letter about "the light of science" dispelling "monkish ignorance" by teaching the fundamental equality of the race of mortals. In Jefferson's memorable formulation, "The mass of mankind has not been born with saddles on their backs, nor a favored few booted and spurred, ready to ride them legitimately by the grace of god."[49] The political effect of the discovery of the immutable laws of motion is to tip the balance toward the common right of humanity as against the divine right of kings.

The 1725 discourse begins by speaking generically of "the arts and sciences" as the avenue of civilization, then quickly pivots to this assertion about the political utility of specifically modern science, before turning at the very end to *belles-lettres* as a promising helpmate to science. Montesquieu highlights the contribution that can be made by literary-minded popularizers—a line of work in which he himself had dabbled in the *Persian Letters*.

However, it seems that Montesquieu fairly early on became concerned about the long-term compatibility of science and freedom. The humbling of kings is all well and good, but the all-out humiliation of humankind is risky. One of his notebook entries laments the alienating effects of Spinoza's reductionist materialism:

> A great genius has promised me that I will die like an insect. He seeks to flatter me with the idea that I am merely a modification of matter. . . . And instead of this immense space that my mind [*esprit*] embraces, he gives me to my own matter and to a space of four or five feet in the universe.
>
> According to him, I am not a being distinguished from another being; he deprives me of everything I had believed most personal. I no longer know where to find again this "me" in which I used to be so interested. I am more lost in the expanse than a particle of water is lost in the sea. Why glory? Why shame? Why this modification which is not oneself? . . . In the universality of substance, there have been—there have passed by without distinction—the lion and the insect, Charlemagne and Chilperic.[50]

Equality can apparently be taken too far. Not only is there an age-old danger that extreme equality can degenerate into political anarchy; there is a profound new threat that emerges from the scientific levelers. Political life depends on discriminations, such as those between great spirits and small, between sagacity and stupidity. As the very title of his magnum opus indicates, Montesquieu values *esprit*—both in its thumotic and noetic senses. Moreover, Montesquieu's psychological acuity makes him sensitive to the requirements of human vanity—"where to find again this 'me'"—especially as that self-preference takes political shape. What conception of human identity is necessary to sustain a public opinion supportive of the practices of political liberty? The dualism of Descartes—whatever its actual status as Descartes's final view (or for that matter as Montesquieu's final view)—may be politically preferable to Spinoza's monism.

In *The Spirit of the Laws*, Montesquieu quite regularly employs imagery associated with the scientific revolution, as when he describes monarchy on the planetary model of centrifugal and centripetal forces. However, these borrowings are decidedly literary. In general, he avoids specifics, moving expeditiously (and ambiguously) past the troubling foundational dilemmas about god, fatality, and law that he floats in the very first pages of the work. Most astonishingly, he nowhere in *The Spirit of the Laws* says anything in praise of science. While there are frequent positive references to the arts, there is only one statement about science, and it is strikingly negative in tone.[51]

In book 4, chapter 8, Montesquieu declares that "the speculative sciences" make men "savage" (in the sense of unsociable). It is possible that he has in mind the more purely contemplative, non-technological science of the ancients, which could turn a thinker into an isolato. Modern science, being dedicated to the relief of man's estate, might not be similarly asocial. However, modern science, for all its practical spin-offs, has certainly not forsworn theoretical speculation (in fact, it is required for the construction of a hypothesis). Our stereotype of the scientist as an odd bird, shy and socially challenged, with unkempt hair and equally wild ideas, is continuing evidence in favor of Montesquieu's conjecture. There is, too, that more serious matter of the possibly brutalizing tendency of the modern atomistic teaching (for scientist and citizen alike).

Montesquieu's other usages of "speculative" shed further, and perhaps surprising, light on his meaning. In 1.2, Montesquieu says that the idea of a creator, as a speculative idea, is a late development; the laws of nature centered on self-preservation come first. In 14.7, Montesquieu attributes the preference for the speculative life to the laziness-inducing effects of a hot climate. Monasticism underwrites this preference and, in the process, impoverishes the common people. In 21.20, Montesquieu blames the "speculations of the scholastics" for "the destruction of commerce" and its accompanying horrors (namely, the torture and expropriation of Jewish merchants).[52] Finally, in 25.4, Montesquieu says that "by the nature of the human understanding," we love "speculatively" all that partakes of severity in religion and morals (his example is priestly celibacy).[53] Judging by these references, theology must rank high among the speculative sciences that render men harsh and savage. Whether Montesquieu was concerned that modern science—through its establishment as the new priesthood, respected but removed from ordinary life—could have similar tendencies is an open question. It is, however, easy enough to interpret the figure of Usbek in the *Persian Letters* as a cautionary tale, inasmuch as Usbek's despotic rule over his harem seems unruffled by his conversion to Western science and enlightenment.

The setting for Montesquieu's lone statement about science in *The Spirit of the Laws* is intriguing. It appears in a chapter that explains the ancient reliance on music "as one of the principles of their politics." According to Montesquieu, because "the spirit of Greek liberty" was grounded in disdain for the arts, agriculture, and commerce, the occupations of citizens were limited to gymnastics and war. These bodily exercises "render men harsh," Montesquieu says, just as the sciences of speculation "render them savage."[54] Since he has not presented the Greeks as given to speculation,[55] the characterization seems gratuitous, although it does allow Montesquieu to argue (perhaps in parody of Aristotle) that music is the mean between these asocial alternatives of too much physicality or too much mentation. Music is also a medium that conjoins body and spirit. Reaching the spirit through the body, it moderates the unsociable effects of overreliance on either. Music, at least in certain modes, has the power to soften and gentle the soul, linking human beings tenderly together.[56]

In one of Montesquieu's own lovely linkages, the verb that he uses to describe the action of music on mores is the very same he uses to describe the action of commerce on mores. Both music and commerce gentle or soften (*adoucir*) those who participate in them. Commerce is the music of modernity. Moreover, on Montesquieu's analysis, commerce is superior because it reliably produces gentle mores, whereas the music of the ancients did nothing more than "curb the effect of the ferocity of the institution."[57] Music was an antidote, perpetually administered to "a society of athletes and fighters," but not a cure.[58] By contrast: "commerce cures destructive prejudices"; "everywhere there are gentle mores, there is commerce and everywhere there is commerce, there are gentle mores"; "the natural effect of commerce is to lead to peace."[59]

Before making these general declarations about the curative power of commerce, Montesquieu, in the immediately preceding chapter, had sketched the character and customs of a free people living under a constitution of separated powers. (That he means the English is clear by his reference to book 11; though here, in 19.27, he never names them, perhaps because his character sketch is meant to apply to "any nation so conceived and so dedicated."[60]) "This nation," he says, "made comfortable by peace and liberty, freed from destructive prejudices, would be inclined to become commercial."[61] This is curious, since the expected direction of causality is reversed. Rather than commerce curing destructive prejudices, commerce, among the English at least, is the consequence of a prior liberation from destructive prejudices. Montesquieu does not disclose the origin of that liberation until the last chapter of part IV, where we learn that Henry VIII "destroyed the monks" (who were a bandit nation[62]), after which "the spirit of commerce and industry became established in England."[63]

It is important to note that while the English were "freed from destructive prejudices," they have not become free of all prejudices. Indeed, Montesquieu concludes his examination of the English character by noting that "each one becomes . . . the slave of the prejudices of his faction."[64] It turns out that some prejudices—as, for instance, political partisanship—may be salutary; they form the gears that drive the dynamic equilibrium of liberty.

While the commercial way of life in England had a despotic catalyst,[65] tossing free-loaders out on their ear is not the only way to develop a society-wide work ethic, as Montesquieu shows in a crucial chapter (19.14) entitled "What Are the Natural Means of Changing the Mores and Manners of a Nation." Here is his rationale for an easier route: "In general, peoples are very attached to their customs; taking their customs from them violently makes them unhappy; therefore, one must not change the customs, but engage the peoples to change themselves."[66] These remarks, paired with the concrete example offered, help us solve the riddle of how Montesquieu can "make it so that human beings . . . cure themselves of their prejudices."[67] Briefly put, the example is of Peter the Great, who introduced the heretofore sequestered noblewomen of his nation to the world of high continental fashion, thereby sweeping away seven centuries of sartorial stagnation and religious orthodoxy. "This sex," Montesquieu explains, "immediately appreciated [more literally, "tasted"] a way of life that strongly flattered their taste, their vanity, and their passions, and they made the men appreciate ["taste"] it."[68] In the chapter "Some Effects of the Sociable Humor," Montesquieu confirms that mores (which have a relatively lasting character) can be transformed into "taste," which is dependent on ever-changing fashion, which in turn "constantly increases the branches of commerce."[69] It all begins from "commerce with women" or "the society of women," since such communicativeness is a specialty of the daughters of Eve.[70]

It is striking that when Montesquieu gives his general description of the commercial cure in book 20, he features the exchange of knowledge (*connaissance*) more than the exchange of tangible goods. According to Montesquieu, commerce causes "the knowledge of the mores of all nations to penetrate everywhere."[71] What happens when diverse peoples encounter one another via communication rather than conquest? Comparisons are made and "great goods"—like the softening of barbarous mores—result.[72] What we today call "best practices," which might have emerged by happenstance (or even violence) in a few places, can now be shared, sparking change elsewhere. Thus, the commercial character, by preserving and fostering openness to innovation, is more in accord with our "flexible being."

Earlier, I said that commerce is the music of modernity. If so, it is music in a new, improvisational and experimental mode. This mode both imitates modern science and inspires the security-minded and comfort-loving character appropriate to enjoy the discoveries of science. We might even say that commerce is the shape in which science (as practical technology) ordinarily manifests itself. Modern peoples even pursue their own version of science, as they participate in the information-sharing society. In keeping with Montesquieu's emphasis on communication, today's social media—Facebook, Twitter, and Pinterest—distribute across the globe the transformative knowledge of comparative mores.[73] It is as he predicted, and apparently welcomed: "As one allows one's spirit to become frivolous, one constantly increases the branches of commerce."[74]

To conclude, let me say just a word about Montesquieu and "nature" (a word he uses in various forms on nearly every page of *The Spirit of the Laws*). Montesquieu usually speaks as a follower of nature, saying things like, "We do nothing better than what we do freely and by following our natural genius."[75] This is not to say that he regards the natural world as particularly friendly to human aspiration. The five books on climate and terrain show him to be acutely aware of despotism-inducing environmental factors. However, he entertains no thought that these elemental forces can be scientifically altered or mastered (as Bacon does in that astonishing account of the doings of Salomon's House[76]). He does, however, think that particular physical causes that are hostile to liberty (such as a hot climate) can be successfully countered by moral causes. Good legislators know how to appeal to elements of human nature—even those that used to be considered discreditable, like vanity and the love of variety—to "incentivize"[77] desirable behavior, like industriousness, thereby creating "happy mixtures" of virtues and vices.[78] Nature is a palette with which one can paint.

Chapter 6

Bacon and Franklin on Religion and Mastery of Nature

Jerry Weinberger

Benjamin Franklin was America's first full-blown Baconian. In April 1732, he published *On Simplicity* in *The Pennsylvania Gazette*.[1] Like all of Franklin's short works, the little piece consisting of ten paragraphs packs a surprising and powerful punch when read with a bit of care. At one point, Franklin says that to find the virtue of simplicity we need not go to the countryside and the "uncorrupted peasants," because it can be found in men of "truest Genius and highest Characters in the Conduct of the world." People who resort to cunning, says Franklin, are just pretenders in matters of policy and business. And then he adds: "Cunning, says my Lord Bacon, is a sinister or crooked wisdom, and Dissimulation but a faint kind of Policy; for it asks a strong wit and a strong Heart, to know when to tell truth and to do it; therefore they are the weaker sort of Politicians, that are the greatest Dissemblers."

This line is a verbatim combination of the opening sentences of Bacon's essays "Of Cunning" and "Of Simulation and Dissimulation."[2] When we return to them, which Franklin surely wants us to do, we discover that Bacon's recommendation of knowing when to tell the truth, and then doing it, clearly means that cunning, simulation, and dissimulation, when well used, require knowing when to tell the truth and when to lie. Moreover, Bacon tells us (on the evidence of Tacitus) that well-used cunning, simulation, and dissimulation are necessary not just in politics and business but also in the arts of life in general. This is a striking conclusion endorsed by a Franklin who, in his *Autobiography*, declared that "truth, sincerity, and integrity" were of the *utmost* importance for happiness in life.[3]

So Franklin knew Bacon's moral thought intimately well. He also knew Bacon's teaching about the conquest of nature. When Bacon wrote *New Atlantis*, he drew up a list called "the wonderful works of nature, chiefly such as benefit mankind." This list was distilled from the description of Salomon's House, the scientific establishment of Bensalem (as the New Atlantis called itself) and was appended to the first (1627) and subsequent publications of *New Atlantis*. It refers not to natural wonders but rather to the things that would become possible by means of modern science as Bacon described it in *Novum organum*: once the latent structures, motions, and spirits of material change were fully disclosed, modern science would fulfill the dreams of the alchemists, the turning of something not gold into gold. Bacon's list includes such wonders to be produced as: prolongation of life; restitution of youth (in some degree); retardation of age; curing of diseases counted incurable; mitigation of pain; increasing of strength and activity; altering of statures; increasing and exalting of the intellectual parts; transformation of bodies into other bodies; making of new species; transplanting of one species into another; instruments of destruction such as war and poison; exhilaration of the spirits and putting them in good disposition; deceptions of the senses; greater pleasure of the senses; artificial materials and cements.[4] This list is astounding in itself. But it also puts the lie to Bacon's claim in "Of The True Greatness of Kingdoms and Estates"—that "no man can by care taking (as the Scripture saith) add a cubit to his stature, in this little model of a man's body"—while asserting that princes and estates can with wise policy add amplitude and greatness to their kingdoms.[5] Jesus apparently did not know what the scientific world would have in store for men's bodies. In "The Wisdom of the Ancients," Bacon says that "natural philosophy proposes to itself, as its noblest work of all, nothing less than the restitution and renovation of things corruptible, and (what is indeed the same thing in a lower degree) the conservation of bodies in the state in which they are, and the retardation of dissolution and putrefaction."[6]

In 1780, Franklin sent a letter to Joseph Priestly in which he said that he regretted being born so soon: "It is impossible to imagine the height to which may be carried, in a thousand years, the power of man over matter. We may perhaps learn to deprive large masses of their gravity, and give them absolute levity, for the sake of easy transport. Agriculture may diminish its labor and double its produce; all diseases may by sure means be prevented or cured, not excepting even that of old age, and our lives lengthened at pleasure even beyond the antediluvian standard."[7] In subsequent letters to other friends, Franklin revised his timeline down from a thousand years to three, then two, and finally to one hundred years. Franklin's Baconian meaning in these letters is pretty clear, even if he does not declare it directly: if old age is a curable disease, then sooner or later, it's so long to the so-called "wages of sin."

As regards morality and mastery of nature, Bacon and Franklin were on very similar wavelengths. They shared a wavelength too about religion. It could well be argued that Bacon's most important popular work, along with the *Essays*, is his posthumous *New Atlantis*. That work describes in great detail the society to produce, and be produced by, the technological project for the mastery of nature. *New Atlantis* tells the story of European sailors who, becalmed and on the verge of starvation, come upon the fantastic island of Bensalem. They are greeted by the people of the island with a certain caution but also with overwhelming humanity. The sick among the sailors are cured by medicines they have never seen, and in a series of interviews with Bensalemite officials, they learn, within the limits of the island's "laws of secrecy," the history of the island, the institution by the wise King Solamona of the scientific establishment, the coming of Christianity to the island, and why the Bensalemites know about the Europeans and yet are unknown to the rest of the world. From a wise Jew named Joabin, they learn about the island's regulation of marriage ("nuptial copulation") and "Adam and Eve's pools," where betrothed Bensalemites get to check out each other's naked bodies, and then, from a "Father" of the scientific establishment, they get an almost full description (because the scientists keep some things secret from the state) of that establishment, "for the good of other nations." Bacon makes it clear that Bensalem differs from the Atlantis related by Plato because Bensalem's science protected the island from the natural disasters (what Plato calls "*divine* revenge") that ravaged the rest of the ancient world.[8]

Even so, religion plays an important and indeed perhaps central role in Bacon's account of the new world to come. Two factors stand out in Bacon's account of Bensalem. First, modern science, in the form of Salomon's House, *precedes* by sixteen hundred years the revelation of Christianity to Bensalem. This reversal of the actual historical facts is odd in itself, but also because Bacon comments elsewhere that the "true religion" had to come into the world in order to free human reason from its temptation to inquire into the mysteries and principles of faith. Since "Salomon's House" was "denominate of the King of the Hebrews," it appears that for the advent of science, the religion of Moses was sufficient. The ancients, whose religion consisted merely of rites and forms (as opposed to confessions and beliefs), could not keep inquiry into nature from drifting into theological and metaphysical discourse, which to Bacon was just so much airy claptrap.[9] The upside of biblical religion for reason is thus its complete separation of reason and faith.[10]

Second, in *New Atlantis* the central religious event is a miracle of original revelation as well as a miracle of tongues. About twenty years after the ascension of Jesus, all the canonical books of the Old and New Testaments, and some other books of the latter that were not yet written, were revealed by the miraculous appearance of a shaft of light pointing to an ark, containing the books, floating in the ocean off the coast of Bensalem. When the books were opened, the Bensalemites, including those who at the time spoke Hebrew, Persian, and Hindi, could read them as if they were written in their own languages. Strikingly, Bacon tells us that a scientist of the House of Salomon happened to be among those who witnessed the event, and he takes it upon himself to verify the miracle as genuine. He prays as follows: "Lord God of heaven and earth, thou hast vouchsafed of thy grace to those of our order, to know thy works of creation, and the secrets of them; and to discern (as far as appertaineth to the generations of men) between divine miracles, works of nature, works of art, and impostures and illusions of all sorts." And then, despite the separation of reason and faith, he testifies to the people that the shaft of light is the miraculous appearance of the finger of God.[11]

From his description of the scientific establishment in Bensalem, we learn from Bacon (in the guise of the fable's narrator) that the scientists have mastered all the techniques of art and illusion and have "so many things truly natural which induce admiration, could in a world of particulars deceive the senses, if we would disguise those things and labour to make them seem more miraculous." We are told that the scientists are forbidden to indulge in such fakery, but at the same time we are told that the scientists take an oath of secrecy for the concealing of things they do not want revealed.[12] According to the timeline of Bacon's story, the scientific establishment was erected in Bensalem almost three hundred years before the birth of Jesus, and the miracle occurred and was verified by the scientist about twenty years after his death and resurrection. So Bacon puts several thoughts into our minds. First, it is possible that the scientists faked the whole miracle. On the other hand, three hundred years of scientific progress could have been enough for the scientists to have verified the miracle as having no natural or artificial cause. That's hard to swallow, however, since at the time the whole story is told to the sailors, the verifying scientist's knowledge was sixteen centuries out of date. Then again, Bacon seems to suggest that in principle *all* the latent structures and processes of nature can be known "as far as appertaineth to the generations of men."[13] So if science were to get to that point, and there is an end to the generations of men who make further progress, we would have to admit that an occurrence unknown to science would have to be miraculous.

On the issue of miraculous revelation, Bacon points us in opposite directions. If the scientists are the masters of deceit and illusion, then most likely the miracle is a fake. Conversely, absolute science would make it possible to know for sure. But that would depend on the scientists being open about everything they know, and they are not. So where does Bacon leave us on the question of science and religion? Shaftesbury wrote that "it was a good fortune in my Lord Bacon's case that he should have escaped being called an atheist or a skeptic when, speaking in a solemn manner of the religious passion, the ground of superstition or enthusiasm (which he also terms a panic), he derives it from an imperfection in the creation, make, or natural constitution of man."[14] If Bacon leaves matters as they stand in *New Atlantis*, it is not clear that Shaftesbury is right about this if, against the faked miracle clue, we turn to the condition of science at its completion and the clear possibility of recognizing a genuine miracle. At the end of the day, Bacon seems to imply that science cannot rule out the possibility that its conquest of the wages of sin is a dangerous undertaking in the eyes of a jealous god.

So we wonder if Bacon had other reasons to take the chance he boldly embraced. The place to look is not so much in his scientific and theoretical writings, but in those popular "self-consuming artifacts," the *Essays*.[15] In the theoretical context to which Shaftesbury refers, Bacon argues that the natural roots of religion are the passions of fear (especially of death) and the natural inability of human beings to separate idle from salutary fears. (This is of course the tack taken by Hobbes, as well.) But in the *Essays*, Bacon makes two completely different and surprising arguments about fear and religion. Death, he says, is not frightening all. Considered as a matter of nature, "there is no passion in the mind of man so weak, but it mates and masters the fear of death; and therefore death is no such terrible enemy when a man hath so many attendants about him that can win the combat of him." Revenge, love, honor, grief, fear, pity, and even boredom either anticipate death or conquer the fear of it. Men commonly die in good spirits or doing what they always do—dissimulating, in the case of Tiberius, or with a jest, as Vespasian did when, sitting on the toilet, he exclaimed that he was becoming a god.[16]

So what is it that makes death so fearsome *when* it is? "Men fear death, as children fear to go in the dark; and as that natural fear in children is increased with tales, so is the other. Certainly, the contemplation of death, as the wages of sin and passage to another world, is holy and religious; but the fear of it, as a tribute due unto nature, is weak." Bacon follows this up by referring to the mixture of vanity and superstition in religious meditations (for example, he refers to "some of the friars' books of mortification"). These friars would remind us, by asking how it feels to torture the end of one finger, how it must feel when the whole body "is corrupted and dissolved." Not to worry, says Bacon in reply: most of the body feels nothing and its most vital parts "are not the quickest of sense." The friars are wrong because death is not very painful. But this move is Three-Card Monte. Those friars were *not* talking about the dissolution of the body, which may well be painless, but its eternal and painful torment in the fires of hell.

Men fear death as they do, despite what Bacon says, because they believe in original sin and punishment in hell. That these doctrines are "holy and religious" does not mean they refer to anything real. And Bacon, in the near-following essay "Of Revenge," gives at least some of his reason for thinking them unreal. Revenge, says Bacon, is "a kind of wild justice; which the more man's nature runs to, the more ought law to weed it out." Revenge "puts the law out of office," and that is reason enough to avoid it. But Bacon then argues that revenge makes no sense. What's past is past, and we've got enough on our plates at present and to come, and "therefore they do but trifle with themselves, that labor in past matters." Fair enough. But then comes the blockbuster: "There is no man doth a wrong for the wrong's sake; but thereby to purchase himself profit, or pleasure, or honor, or the like. Therefore why should I be angry with a man for loving himself better than me? And if any man should do wrong merely out of ill-nature, why, yet it is but like the thorn or briar, which prick and scratch, because they can do no other."[17] Bacon wrote no essay "Of Justice," but if he had, these last arguments would apply to it as well. And the most powerful of them is not that we should not blame another for loving himself better than ourselves (or that we are naturally selfish). It is rather the clear implication that when we do wrong, we think we are doing *right*, because nobody does a wrong for the wrong's sake. And if someone in fact *does* a wrong for the wrong's sake (perhaps out of "sin" or "evil"), that person is no different from a thorn that cannot do otherwise. If a thorn scratches, you might just pull it off; but you would not blame it as well.

Bacon is asking us to imagine that Person A makes wicked choice "B" knowing that "C" is the right thing to do. What kind of person makes such a wicked choice? A wicked person. How did A become wicked? By rejecting what he knew to be good and making the wicked choice, we say. But to make that wicked choice knowingly, A had already to be wicked. Where, then is the choice for wickedness? If in blaming we appeal to wickedness as opposed to ignorance (thinking the wrong thing to be right), then we are stuck with saying that a wicked person is one who is wicked, and no more. It is a condition, not a choice, of a person who cannot do otherwise, just like Bacon's thorn. We would perhaps kill that person as we would a poisonous snake, but it would be as absurd to blame the person as it would be to blame the snake.[18] Fear is not the source of faith; that job falls to our vivid moral intuitions, and these intuitions are hopelessly confused, says Bacon. That said, for most people, including the Bensalemites, this confusion does nothing to cause these intuitions to lose their power. At any rate, it is quite possible that Bacon did not fear taking on the wages of sin, because the idea of sin, original or otherwise, made no sense to him.

In 1735, Benjamin Franklin weighed into the Presbyterian controversies roiling Philadelphia in favor of the good-works preachers and against the orthodox Presbyterian establishment. In one pamphlet in support of the Reverend Mr. Hemphill (who was later exposed as a plagiarist), Franklin said the doctrine of original sin is an absurd "bugbear set up by Priests (whether *Popish* or *Presbyterian* I know not) to fright and scare an unthinking populace out of their senses, and inspire them with terror, to answer the little selfish ends of the inventors and propagators." It is monstrous, says Franklin, to say that a man could be punished for the guilt of another, and to believe this, as the Presbyterians do, is to make God arbitrary, unjust, and cruel.[19]

Despite this vehemence, Franklin, who with Bacon had the wages of sin in his scientific sights, was much more sympathetic to religion than were the major heroes of the Enlightenment. Franklin was, in his way, an Enlightenment critic of Enlightenment rationalism. The famous *Autobiography* is not a priggish and utilitarian story of how to become rich and famous. It is actually an extraordinary story of how Franklin, born a good boy and into a God-fearing home, was corrupted by the thinkers of the Enlightenment. They made him into a thorough skeptic and led him to write a metaphysical tract that denied the existence of virtue and vice, and so too, divine providence (this he did in London at the age of nineteen, and it got him introduced to Bernard Mandeville). After that he harmed his friends, some of whom then harmed him. It was only a subsequent moral and then religious self-redemption, he tells us, that saved him from catching syphilis. He gave up on *any* metaphysical answer (either pro or con) to the question of God and particular providence, got married, and then became the First American and moral model for a new and rising people.[20]

The story of self-redemption is untrue. But, even so, Franklin followed a double track afterward. On the one hand, he made up and published hilarious and sometimes grotesque stories about religious enthusiasm and madness. These stories make the House of the Lord loonier than an insane asylum. But on the other hand, Franklin also *countered* the canards with the astonishing and double-sided 1730 "Letter of the Drum" and the reply, three days later, from one Philoclerus, both in the *Pennsylvania Gazette*.[21] In the first piece, Franklin describes the absurd query of man who, from the influence of "Spinosists and Hobbists," had become a convinced and "impious free-thinker" and so no longer feared spirits and witches. Alas, however, a Reverend Gentleman rekindled his faith and fear by recounting a miracle that occurred while the Reverend and his fellow clergy, in town to plan a campaign against atheism, were in bed. The Devil banged a drum that no one else could hear and then grabbed one of the preachers by the toe. The Reverend would not make all this up, the poor fellow says, so he now again sleeps with his curtains tied together to keep out the spooks. What should he do, he asks the *Gazette* editor: believe the Reverend or not?

In the second piece, Franklin, as Philoclerus, admonishes Franklin not to make fun of religion, if only because of all the good it does for society. But then he makes the argument that it is not so easy to disprove the existence of spirits: on the one hand, Hobbes and his followers come to materialistic conclusions because they begin with materialistic assumptions (there are no spooks because there are no spirits); and on the other hand, the "Hobbists" could *explain* the existence of spirits and miraculous revelation materialistically.[22] Either way, those "Spinosists and Hobbists" are on much shakier ground than their skeptical followers think. Like Bacon, it seems, Franklin came to think that neither a metaphysical system nor materialistic science could settle the issue of providence and divine revelation. His follow-up take on the issue is very close to Bacon's as well.

As a much older man, Franklin served on the committee to write the Declaration of Independence. A later editor and biographer of Franklin, Albert Henry Smyth, wrote in 1905 that Jefferson explained that Franklin was not asked to draft the document because "he could not have refrained from putting a joke into it."[23] No one knows if this is really true, but it could be. There is nothing that Franklin failed to ridicule, often hilariously and always with great subtlety beneath burlesque exteriors. Still, the Declaration of Independence? Franklin was a gifted propagandist and knew when to hold his tongue. There is, however, a simple explanation for the Declaration episode if indeed it really happened. Franklin did not think the doctrine of natural right made even the slightest bit of sense.

In 1732, Franklin presented to his intellectual club a set of proposals to be discussed. One proposal asked, is it "justifiable to put private men to death for the sake of public safety or tranquility, who have committed no crime . . . as in the case of the plague to stop infection, or as in the case of the Welshmen here executed?"[24] Franklin furthermore asks, "If the sovereign power attempts to deprive a subject of his right, (or which is the same thing, of what he thinks his right) is it justifiable in him to resist if he is able?" If a right is nothing more than what any man *thinks* is his right, then surely there is no binding natural right; no natural right that other men are duty-bound to respect. Moreover, Franklin did not reject the principle of natural right dogmatically. He came to his conclusion because he did not think that any human being deserves anything (whether reward or punishment): not from nature, God, nor other men. Franklin worked out this position by considering our variety of moral intuitions and points of view in short, often comical writings, including a "Socratic Dialogue" (*A Man of Sense*), written in 1734–1735, starring Socrates and a very stupid man named Crito.[25] One especially representative piece is the 1757 "A Letter from Father Abraham, to His Beloved Son," in which absolute moral vice is presented in the image of a little boy (the beloved son) who, thinking them to be good for himself and others, farts away (small acts of dishonesty) until he winds up with a "Sir-Reverence" (full-blown vice) in his breeches.[26] In a different take on our sense of deserving and reward, Franklin depicts men driven to naked madness when, having only their pure deserving but no corresponding happiness from their devotion to Jesus, they try then to rid themselves of that pure deserving, because it is the highest *reward*, for themselves, for their pure devotion. (Franklin was as far from Kant as it is possible to be.)[27]

The last of these pieces, written just weeks before Franklin died, was about slavery. That took guts for someone who rejected the doctrine of natural right *and* had one foot in the grave. The 1790 piece is in the form of a speech by Sidi Mehmet Ibrahim, a member of the Divan of Algiers, which turns out to be an exact replica of Georgia congressman James Jackson's speech against congressional interference in the slave trade.[28] But it also turns out that Ibrahim's speech is a defense of the enslavement of "Christian dogs" against the radical "Purist" Muslims who, bearing a great burden of sins and hoping by good works to be excused damnation, want to outlaw the practice. In the course of this subtle and complicated hoax, divine law serves both sides of the slavery question, and moral duty is also invoked on both sides. As Franklin puts one pro-slavery argument in Sidi's mouth—that the Muslims need enslaving piracy to provide the "commodities their [European] countries produce, which are so necessary for us"—Franklin forces us to see that, as pirates, they settle for less than the Europeans produce (unless they can steal most of it, which is hardly possible) and will have to rely on the fortunes of war, which the Europeans will win because they make things and the Muslims, made lazy by slavery, do not. Franklin elsewhere reminds us that all the glories of the ancient world were based on slavery: it is mastery of nature that frees us from the awful necessity, which for Franklin included the slavery of degrading but free labor. That is why, as he wrote to John Bartram, from Italian travels he would rather see the recipe for Parmesan cheese than any old inscription on a stone.[29]

Franklin makes the case against slavery on prudential rather than moral grounds: it is easier to convince men that slavery is imprudent than to convince them that their moral principles—their conceptions of right and deserving—are incoherent. Franklin's position on slavery at the end of his life was something like this: blacks and whites are equally intelligent, so there is no claim to be made that slavery is good for either one; and, in fact to the contrary, it degrades the enslaved and makes the masters *indignantly* cruel and lazy at the same time. Were there a real necessity, say in a state of nature, for Franklin and another man to be either slave or master, Franklin would say, "Better you the slave than me." But since he did not think either he or the other man had anything coming to him by "desert," the master had better sleep with one eye open because the slave has no reason to stick around (it would all depend on how bad things got), or not to turn the tables on the master when he can, or kill him if necessary.

Toward the end of his life, Franklin wrote a letter to his son William, who had remained a Tory and thus took up arms against his father "in a cause wherein my [Ben's] good fame, fortune, and life were all at stake."[30] William had claimed that his "duty" to his king and regard for his country required him to take this stand. To this Franklin replied: "I ought not to blame you for differing in sentiment with me in public affairs. We are men, all subject to errors. Our opinions are not in our own power; they are formed by and governed much by circumstances, that are often as inexplicable as they are irresistible. Your situation was such that few would have censured your remaining neuter, though there are natural duties which precede political ones, and cannot be extinguished by them." This argument about our opinions—that they are conditions we suffer rather than choices we make—occurs often in Franklin's writing and always to the same end: when we do anything, we do so in the grip of an opinion that it is right and good, and that goes for William as well as Sidi Mehmet Ibrahim. Franklin chides his son for neglecting his "natural duty" to his father, by which he has to mean the ties of the heart. But even if there were such a natural duty of the heart, the argument Franklin makes about William's political opinions would apply to it as well.

Franklin was not a believer, but he always touted a latitudinarian "creed" that included God's particular providence, rewards, and punishments, in this life or the next, for the good and bad things we do. He could not have believed this creed himself, convinced as he was that no one deserves anything from others. But he said frequently that if men are as bad as they are with religion, imagine what they would be like without it. There was, however, a deeper reason why he touted religion, and it is very similar to the reason that Bacon has religion and revelation at the center of *New Atlantis*. Both Bacon and Franklin, unlike Hobbes, thought that religion springs not from fear, but from the ineradicable human hope that we will get what we deserve in life. The Bensalemites have little or even nothing to fear, but even so, they look like a doped-up herd of sheep and exhibit a grinning but robotic and even creepy religiosity. Perhaps that is what is necessary to keep them from one another's throats, because whatever plenty they might have, they are *always* capable thinking, erroneously, that they deserve, especially as individuals, still more and better.[31]

Franklin, much more than Bacon, had a profound understanding of how stupid and foolish human beings can be, and always will be, so long as they believe they deserve things and have them coming from others. They will always so believe, Franklin thought, and so no degree of the mastery of nature will overcome religious enthusiasm or, for that matter, its secular counterpart: political partisanship. Franklin's moral compass proceeded from the negative, not the positive light of duty. In his comedies and doodles he shows us how confused we are as deserving individuals and how that confusion leads to self-torment and, especially, to the moral indignation at the root of all the awful things human beings do to each other. Scratch a crook and you'll find somebody getting even. When a tyrant thinks that "might makes right," he thinks he is in the *right* because his might makes him so. Political partisans of every stripe think they deserve to win and rule. On the question of slavery, he makes us ask if any slaveholder in the South, whether kind or cruel, looked in the mirror every morning to exclaim, "Now there's the face of evil!"

At the height of the Revolution, Franklin published an astonishing hoax reporting the discovery of a grotesque cargo of white scalps taken by Indians and being shipped to England as tokens of affection to the king, Parliament, and the bishops.[32] But in the shipment was also found a pathetic transcript of a speech by Chief Conejogatchie, declaring the scalps to be tokens of his grateful affection and faithfulness to the king. Then comes a plea from the chief to the king for help against his increasingly powerful "enemies" (the Americans), who have driven them out of their country, and who have "also got great sharp claws." The butchery has cut both ways, and Franklin knew early on that the Indians would continue to get the worst of it down the road.[33] Of those Indians, he said not even a year later: "Savages we call them, because their manners differ from ours, which we think the perfection of civility; they think the same of theirs."[34] It was not the justice of one side, thought Franklin, that led it to win and the other to lose. After the war, Franklin wrote to Priestly that the whole enterprise had resulted from stupidity and angry indignation and was a senseless waste of blood and resources.

Franklin really did think charity is a good thing and, with Bacon, that the modern scientific project is its most powerful form. He thought that project would be most likely to flourish in America. That said, Franklin, following Bacon, never had high hopes for people becoming reasonable creatures, because he did not think human beings would ever stop thinking that something—their dignity, virtue, or deeds, who or what they are—makes them worthy of a free lunch from God or other men. Franklin thought the mastery of nature a very good thing, but he did not think it would eliminate the need for wise and pragmatic political engineering: the inevitable ranks of religious enthusiasts and political partisans would always require manipulation by smart opportunists like him.

PART II

Ancient Alternatives and Anticipations

Chapter 7

On the Supremacy of Contemplation in Aristotle and Plato

Robert C. Bartlett

> When antitheological passion induced a thinker to take the extreme step of questioning the supremacy of contemplation, political philosophy broke with the classical tradition, and especially with Aristotle, and took on an entirely new character.
>
> —Leo Strauss, "Marsilius of Padua"

The purpose of this volume, I take it, is to begin to understand the concern characteristic of the modern philosopher-scientists to effect "the mastery and possession of nature" or the "conquest" of nature—to torture or otherwise compel it to give up its secrets—in contrast to that of the ancient philosopher-scientists who were evidently content just to look on or observe or "contemplate" nature. The modern thinkers, that is, seem united in their hope to put to use the necessities of nature, once discovered, for "the relief of man's estate," as Francis Bacon put it,[1] whereas the ancients declined to pursue this possibility. And they declined to pursue it for reasons other than that they had failed to conceive of it. As Xenophon reports, at any rate, Socrates examined whether it was the case that, just as those who understand the human things hold that they will do whatever they wish with what they know, so also those who investigate the divine things believe that, once they come to know the necessities by which each thing comes into being, they will make, whenever they wish, winds and waters and seasons and whatever else they may need of such things. Or, Socrates continued, is it rather the case that those who investigate the divine things hope for no such thing at all but are satisfied instead just to know in what way (or through what) each of the things of this sort comes into being (*Memorabilia* 1.1.15)?

At least in comparison to their modern counterparts, the ancient philosophers must be said to belong to those who sought only to know the necessities and not to do or make things with that knowledge: Thales's cornering of the olive press market, on the basis of his knowledge of astronomy, and Aristotle's anticipation of looms and plectrums that can move of their own accord, surely do not constitute exceptions to this (*Politics* 1259a6–19; 1253b33–1254a1). And if Empedocles may well have permitted Pausanias, his student, to come to know the drugs then in existence (*hossa gegasi*) that defend against ills and old age, there is no reason to think that he ever made good on his boast to bring about, for Pausanias alone, the capacity to manipulate winds and rains or to restore the strength of a man dead and gone in Hades—this last capacity presumably much diminishing the worth of any preventative drug (Empedocles fr. 111).

It is beyond my abilities to explain the characteristic stance of Platonic-Aristotelian philosophy toward nature. But two preliminary points suggest themselves. First, if (as has been suggested) the modern philosophers sought to remake the world in order to respond to the challenge to all scientific orientations posed by biblical revelation in general and by miraculous intervention in the "natural" order in particular, it would seem that the classical thinkers lacked that incentive, and not because they did not face a comparable challenge. They evidently dealt with it in a different manner. Second, if it is true that the mastery of nature, as distinguished from the contemplation of it, would require the popularization of philosophy, Plato and Aristotle give evidence in their moral and political writings of having thought that the popularization of philosophy would require such changes to both philosophy and the political community as would do grave harm to each. At any rate, Socrates clearly opposed the new or newfangled wisdom, represented in the dialogues of Plato by the sophists Protagoras and Hippias, according to which the wise may speak with far greater frankness than "the ancients" had realized (*Theaetetus* 180c7–d3; consider also *Greater Hippias* 281c3–283b4). Where many of "the ancients" had taught covertly that all is in motion, by saying, for example, that the two aqueous divinities Oceanus and Tethys are responsible for the coming into being of all things, the later thinkers were so confident in their greater wisdom that they taught openly, even to shoemakers, that all is in motion. What is perhaps more, these later thinkers supposed that the shoemakers would even honor them for their wisdom—and this despite the fact that the teaching concerning the ubiquity of motion proves to destroy the world of concern to shoemakers (and not only to shoemakers).

The outspokenness of a Protagoras, however, appears to be circumspection itself when one compares it to the speech of *our* moderns. For even Protagoras was concerned above all with "virtue" as he understood the term (the correct union of courage and wisdom); and this distinguishes him, to say nothing of Plato and Aristotle, from his modern counterparts—not excepting Machiavelli, at least in the sense that his treatment of precisely "virtue" made possible or necessary the subsequent abandonment of the term. For we read in Machiavelli such statements as these: "If one considers everything well, one will find something appears to be virtue, which if pursued would be one's ruin, and something else appears to be vice, which if pursued results in one's security and well-being"; "it is necessary to a prince, if he wants to maintain himself, to learn to be able not to be good, and to use this and not use it according to necessity" (*Prince* ch. 15). Necessity thus demands that one look to one's own security and well-being above all, which includes not least maintaining oneself in power. As a result, it becomes permissible to speak of a man's "inhuman cruelty" together with his "other virtues." It fell to Thomas Hobbes to translate the point into more acceptable language: "there is no such *Finis ultimis* (utmost aim) nor *Summum Bonum* (greatest good) as is spoken of in the books of the old moral philosophers" (*Leviathan* ch. 11).

I propose to sketch the argument of Plato and especially Aristotle concerning the centrality to human life of the concern for "virtue," above all for contemplative virtue, in its relation to the *Summum Bonum*, or happiness. As it seems to me, the classical analysis of virtue and happiness is alive to the kinds of criticisms that would come to be leveled against it by the modern thinkers—and yet Aristotle, like Plato, insisted on virtue's centrality nonetheless. Why did they do so? What price did they think we would pay were we to abandon, or attempt to abandon, that concern for a more "realistic" one?

Aristotle on the Problem of Happiness

According to Aristotle's famous formula, happiness is "an activity of soul in accord with virtue." What we prove to mean by "happiness" is the possession of a good that renders our lives "self-sufficient and in need of nothing," nothing either better or higher or somehow more complete than the good that is precisely happiness (NE 1097b14–15). Now Aristotle insists that, in order to be happy, we must of course have available to ourselves the goods of the soul, or the virtues. But necessary too are the goods of the body (a measure of health, for example) and "external goods"—at least some wealth, some friends, and so on. And since these bodily and external goods are finally dependent on chance, according to Aristotle, so too, it would seem, is our happiness (consider NE 1098a18–20; 1099a31–b11; 1100b4–7 and context). This is *the* problem of happiness as Aristotle first sketches it. If we come to know, to feel in our bones, the exposed character of our hope for happiness, we naturally enough rebel against that exposedness. For example, as Aristotle notes, some people are inclined simply to equate happiness with chance—we are according to them but the plaything of chance—which amounts in effect to a denial of the possibility of happiness, since we divine it to be something stable and lasting. Others, quite to the contrary, are inclined to equate happiness with moral virtue. And this dedication to moral virtue as the only thing we need to be happy can amount to fanaticism: the man being tortured on the rack, provided he is morally good, is happy, some will argue (consider NE 1153b19–21 and 1095b30–1096a2). Aristotle will certainly explore the possibility or hope, which every morally serious person will surely have at some point, that happiness may arise solely from acting in accord with moral virtue—only to return to the necessity of external goods (see the crucial turn at NE 1099a30 and following). Yet Aristotle also suggests that entrusting happiness, "the greatest and noblest thing," to chance would be "excessively discordant" or harsh (NE 1099b24–25). What then can be done?

Another response to the problem of happiness is to posit both an afterlife and the existence of gods who are willing and able to provide and protect the happiness of human beings, in accord with their merits: the just man being tortured on the rack is of course not happy in the midst of his agony, but he may hope for a perfect compensatory happiness in the next life if he has acted nobly in this one. So it is that Aristotle immediately introduces, in Book I of the *Ethics*, the question of whether happiness may be due to "divine allotment," as a gift from the gods, just as he will soon suggest that the dead must be aware of what happens to their living descendants—though not in such a way as to "make the happy unhappy or anything else of that sort." There is, then, a happiness available to us after death that nothing can fundamentally disturb.

But it has to be said that these comforting views, of happiness as god-given and of a happy afterlife, are asserted but not proved by Aristotle. The first view rests on a conditional clause ("*if* there is in fact anything that is a gift of the gods to human beings"), just as the assertion of a life after death depends on the observation that to deny it would be "excessively opposed to what is dear and contrary to the opinions held." Yet Aristotle also says that, precisely if happiness is an activity, it is "absurd" to say that the dead can be happy: some harsh and paradoxical things may nonetheless be true.

The first book of the *Ethics* is meant not to solve the problem of happiness but to clarify it. That is, it sets forth our hopes for happiness and sketches the most serious alternatives for overcoming the power of chance, be it a zealous dedication to moral excellence or a pious devotion to providential gods. When Aristotle turns to his detailed elaboration of the eleven moral virtues, the question of happiness all but vanishes, together with piety as a virtue. But what precisely does it mean to drop piety from the list of the moral virtues, the virtues of character, and so to demote it in the lives of his thoughtful readers? It means that Aristotle wishes to encourage in them the conviction that dedication to virtue is sufficient, if not quite to guarantee the happiness of the virtuous, then to secure the most upright conduct and hence the greatest nobility in even or precisely the most difficult circumstances, when the power of chance cannot be averted. Virtuous Priam, king of Troy, suffered terribly in the end, as Aristotle notes, and yet it was just then that his nobility most "shone through." Aristotle wishes to encourage the view, on the part of his "noble and good" reader, that virtuous activity is often equivalent to happiness—at any rate, that it "exercises authoritative control" over our lives (NE 1100b9–11, 33–34). As for those undeniable circumstances in which misfortune thwarts our hopes for happiness, there is yet great solace in the certainty that we can still act as nobly as possible, in the manner of Priam. Perhaps the very sacrifice of our hopes for happiness that stem from noble action, while we remain dedicated to noble action, is itself of sublime nobility. Surely Aristotle wishes to encourage such virtuous resolve, or toughness, in his audience and so to encourage them to rely less on the gods—to be less like Nicias when the chips are down than like Demosthenes, as Thucydides presents them (*War of the Peloponnesians and Athenians* VII.50, 69, 77, 82–85; consider also Aristophanes *Knights* 30–36 and context). It is with this sobering aim in mind that Aristotle sometimes plainly denigrates the Olympian gods in favor of the nameless "god" who is without external actions.

Yet does such resoluteness really amount to the greatest possible clarity of mind? I believe that Aristotle's answer is "no." When he returns, at the end of the *Ethics*, to the question of happiness, his treatment of it differs in some important respects from the earlier one. Above all, only now does Aristotle reveal that, of the two distinct kinds of virtue, moral and contemplative, moral virtue by itself is incomplete. Such happiness as is available to human beings is bound up with activity in accord with contemplative virtue only. Moral virtue arises through habituation rather than education (NE 1103a14–23); it is a necessary means to make communal life possible and is tied to the regulation of the passions and to the needs of the body (consider NE 1178b9–16; 1178b33–35).

Much and perhaps everything is thus made to depend on contemplative virtue. Accordingly, in the climax of the *Ethics*, Aristotle offers a series of arguments to establish the coexistence of contemplation and happiness. In this way, whatever hopes for happiness the dedication to moral virtue may have aroused before in us, we must now transfer to contemplative virtue. But precisely how does the activity of contemplation—understanding, say, the nature of motion—solve the problem of our necessary dependence on the goods of fortune, a problem of which Aristotle here reminds us (NE 1178b33 and following)? These are Aristotle's chief arguments: contemplative virtue is the full development of our intellect or mind, which either is divine or is the most divine thing in us; the activity of contemplation is the most continuous one of which we are capable; and, happiness not being without pleasure, contemplation offers pleasures of great stability and purity. Above all, contemplation or thinking, in addition to being cherished for its own sake rather than for any goods that may come from it, also offers the greatest possible self-sufficiency for a human being, in the sense that we least of all need external goods (or even other human beings) to think. Aristotle concludes this long series of arguments connecting contemplation with happiness in this way: "If all this is so, then this activity would constitute the complete happiness of a human being." But he adds immediately this proviso: "provided, that is, that it goes together with a complete span of life, for there is nothing incomplete in what belongs to happiness." And, lest we miss this jarring conclusion, Aristotle states "but a life of this sort would exceed what is human" (NE 1177b24–28).

If our hopes for happiness cannot be realized as we wish them to be through any human means, not even by thinking of the most exalted kind, the possibility of divine providence understandably resurfaces. We have returned, on the level of contemplative virtue, to the key problem that plagued moral virtue, too. And here Aristotle offers two distinct paths to pursue. According to the first, the intellect is the best or most divine thing in us, and to live in such a way as to perfect it *is* to enjoy a divine happiness—to the extent possible. "Happiness, then, is . . . coextensive with contemplation; and the more contemplation is possible for some, the happier they are—not accidentally but in reference to the contemplation in question. For it is in itself honorable. As a result, happiness would be a certain contemplation" (NE 1178b28–32). Or, as he had put it somewhat more accurately shortly before, the life that accords with the intellect would be "the happiest" one; that is, the happiest available to human beings (NE 1178b23). This, it is true, "solves" the problem of happiness in part by limiting the extent of our dependence on the things of chance, for, to repeat, the life of contemplative virtue is most self-sufficient. We remember that *the* good we seek would render our lives "self-sufficient and in need of nothing." But no human life is fully self-sufficient, and so Aristotle's argument thus far grapples with the problem of chance also by permitting us to see the necessary or natural limits attending our lives and so encouraging us to accept those limits. To accept the supreme goodness of perfecting the human intellect does not solve the problem of chance in the sense of making it go away; to understand the world as it is and ourselves as we are is to see that our original or pre-philosophic hope for a complete happiness is not realizable.

But this is not Aristotle's last word. According to the second path he here sketches, those who actively seek to perfect their intellect "also seem to be dearest to the gods. For if there is a certain care for human beings on the part of gods, as in fact there is held to be, it would be most reasonable for gods to delight in what is best and most akin to them—this would be the intellect—and to benefit in return those who cherish this above all and honor it, on the grounds that these latter are caring for what is dear to gods as well as acting correctly and nobly" (NE 1179a23–29). In other words, Aristotle's last word on the subject in the *Ethics* is that there may be a sort of divine providence after all, one that actively rewards not moral virtue but theoretical excellence. This does not eradicate the moral hope for a complete happiness but shifts its focus.

Aristotle on Virtue

Here a few more general remarks may be in order. A defining feature of Aristotle's political philosophy is its invention or discovery of "moral virtue," to which he encourages the better part of his general audience—the "serious" or "noble and good"—to dedicate themselves. Aristotle may construct moral virtue out of elements of the ordinary understanding of human excellence as he found it—what Plato's Socrates had harshly called political or "vulgar" virtue—but that very construction constitutes a refinement meant to bring greater order, reason, and contentment to the life of action, even as it makes that life less hostile to the life of the mind. Aristotle, too, grants that courage or manliness is a real virtue, for example, and he treats it first—not, however, because it is the best or most important of the virtues, but because it proves to be the least or lowest of them, it being bound up with a "non-rational" part of the soul (NE 1117b23–24; consider also Plato *Laws* 625c9–626b4, 630a7–b2, 631c7–d1). Moreover, in his exposition of the eleven moral virtues, Aristotle refuses to deduce them from the demands of community or indeed from anything supposedly higher than they: he presents the moral virtues as being choiceworthy for their own sakes, or at any rate "for the sake of what is noble" (NE 1115b12–13). The life of moral virtue thus occupies a middle ground between the complete immersion in political life and the entirely apolitical or inactive life of "philosophy" so conceived; the life of moral virtue is compatible, to say the least, with decent citizenship but understands itself to have a foundation different from service to any particular community, however good. While Aristotle admits or insists that this life falls short of the highest peak of which human beings are capable—the attainment of contemplative virtue—it soars above any ordinary or unreflective life because it is elevated by the greater reasonableness of its concerns and by the worth of what it has been taught by Aristotle to admire.

At the same time, the distance by which moral virtue falls short of what is highest must be measurable in the lives of those who hold to it alone. Above all, the life of moral virtue ultimately fails to provide us with what we most wish for: a happiness deserving of the name. About this failing Aristotle is remarkably explicit. As a partial remedy or response, he sketches in the *Politics* the peculiar "catharsis," or purgation, effected through music, of certain passions, above all those of "pity and fear."[2] But, as we have seen, Aristotle also encourages the more educated and refined to admire the seriousness of those, like King Priam, whose dedication to nobility remains even when they can have no reasonable hope for happiness. After all, is not moral virtue to be practiced for its own sake, without regard to one's own happiness? The structure of the *Ethics* as a whole raises just this question, for it begins and ends with discussions of happiness, between which is the account of the moral virtues that scarcely mentions the word "happiness."

Aristotle thus reproduces, he respects, a crucial fact about us: our very great longing to be happy, to possess for ourselves the greatest possible good, is accompanied by our deep attraction to doing the right thing for the right reason, even (as it may be) at the cost of our own happiness. We long to be happy, of course, but we are deeply attracted also to acting "for the sake of what is noble," come what may for ourselves. We are thus attracted both to the attainment of our own happiness and to the sacrifice of it. This mystery of the human soul suggests one reason why we may seek with all our hearts for a god and why Aristotle insists on keeping a place, albeit a restricted place ("fifth and first"), for priests and their concerns in the best regime: only the intervention of a god could somehow reconcile the sacrifice of what appears to be our happiness here and now with the hope to attain a true and hence lasting happiness eventually.

Plato and the Rebellion Against Mortality

We may now cast a brief glance at Plato. At a dramatic moment in the *Theaetetus*—the dialogue that addresses the question "what is science?"—Socrates cautions his elderly interlocutor, Theodorus, against thinking that bad things or ills (*ta kaka*) could ever perish from this world altogether, "for it is necessary that something always be in opposition to the good" (176a5–6). In fact, such ills "prowl about the mortal nature and this place here of necessity" (176a7–8). Thus speaking twice of "necessity" in short compass, Socrates indicates that the goods we "mortals" can know in this world must be attended by something bad. The good that life itself is must of necessity be shadowed by the bad that is its opposite. Now it was surely the case that Socrates, a human being who hated "the lie in the soul," accepted the world as he understood it has to be; he must have accepted in particular the necessity of death for "the mortal nature." And truly to accept such a thing—would this not amount to ceasing to rebel against it, either in one's wish or (still less) in one's deed? The harshness of this may be compensated for, to a degree, by the fact that precisely necessity, at any rate knowable necessity, makes possible understanding. The same necessity, dictating that all that comes into being must also perish, is at the same time the foundation of such wisdom as is available to human beings. To wish away necessity is to wish away human wisdom, too. And this must be impossible for a "philosopher," a lover of or friend to wisdom.

It is not a little surprising, then, that in the very context under consideration, Socrates informs his interlocutor that there *is* a place free of all evils, namely, where the gods dwell, and so it is that "one ought to try to flee from this place here to that place there as quickly as possible," a flight that is constituted by "assimilation to god to the extent possible" (176a7–b2). Such assimilation or imitation, in turn, consists in the greatest possible practice of justice and piety, together with prudence, here and now. Socrates thus encourages Theodorus to judge this world, with its unavoidable ills, in the light of another world free of all ills; Socrates encourages Theodorus to live his life here in dedication to justice and piety, together with prudence, always with one eye on the other place purified of all ills and hence (one must assume) of death itself. Yet Socrates's argument proves to be ambiguous in a manner comparable to the ambiguity of the providence said by Aristotle to attend contemplation. For Socrates also suggests that the practice of injustice here will prevent one from *entering* the abode of the gods, with its unadulterated goods, and so will condemn one to live here below; the just, then, will know a truly divine happiness in another and very different realm from this one after they have met their end (177a3–8). In brief, Socrates sketches two different responses to the problem of justice: to strive to imitate here and now a god-like justice that is as such accompanied by understanding or prudence (176b3–177a3); or to gain entrance to what one may loosely call a paradise very different from the mundane.

To be sure, Socrates doubts that the so-called "clever," who pride themselves on their successful injustice, will accept this second possibility. But if they were to come to accept the former argument and—therefore—to reject the latter, then they would reconcile themselves to living here and now with the very real goods available to us, certainly, but also with the all-too-real ills that necessarily shadow them. To make this choice in full awareness of its consequences would be to live here and now, having accepted that necessity. No wonder, then, that the "clever," who understand themselves to be "real men," are repeatedly addressed by Socrates in terms of the "manliness" or courage required for them to see the world and themselves as it and they are (176c4, 177b3 and b4). Manliness is required, in other words, in order that one not attempt to "flee" this world as it is (compare *pheugein* and *phugē* at 176b1 with *phugein* at 177b4). Yet Socrates tries to do justice to what he presents as facts about human beings: that we are naturally given to rebel against our mortality or to flee this world of harsh necessity; and that we are deeply moved by the prospect of a life whose center is a dedication to god-like justice. And any science of human beings would have to take these facts, if they are facts, fully into account or risk distorting the very phenomena the science is meant to explain.

Conclusion

Aristotle's treatment of virtue and the *Summum Bonum* is still relevant for us because the concerns that guided Aristotle's caution or moderation retain their importance in the economy of human life and work their subterranean influence, acknowledged or not. For it was above all the problem of happiness, of the *Summum Bonum*, that guided Aristotle's exposition of virtue in the *Ethics*, in the first place the threat that unconquerable chance poses to our hope for happiness. To understand or sense that all we most care about is exposed to the meaningless sway of chance, with no power above us to thwart it, is to experience a kind of free-fall. Of course, the astounding labors of the greatest modern philosopher-scientists were explicitly devoted to conquering not only nature but chance, too—a goal as yet unachieved. No wonder that ours is an age rich in anxieties and neuroses.

But the problem of happiness goes beyond even the realm of chance. For at one point Aristotle wonders whether those who act in accord with complete virtue and enjoy sufficient external goods over the course of a long life—those, in brief, who are both good and favored by fortune—may thus be deemed "happy." He replies only that "those among the living who have and will have available to them the things stated are blessed—but blessed human beings" (NE 1101a19–21). It falls to human beings, to mortals, to rebel against being such and to wish instead to be immortal (consider NE 1178b7–9 and 1177b26–1178a2). The study of the political philosophy of Aristotle and Plato turns us toward (or returns us to) the question of the status of the longing for eternity in the human soul. If the world is such that we cannot have what we most desire, philosophic resignation may suggest itself where thoughtful piety does not avail.

Evidently dissatisfied with these alternatives, and perhaps fueled by "antitheological ire" or "passion," the modern successors to Plato and Aristotle experimented with the thought that both politics and the human soul can be rescued from the forces distorting them, thereby restoring each to a truly natural health. Plato and Aristotle would hold, I believe, that the moderns have misunderstood the human soul and—therefore—the community best for it. At any rate, the political philosophy of Aristotle and Plato may help us gain a critical distance on the modern experiment, with its undeniable benefits and what now appear to be its considerable costs.

Among some today, I am told, the desire to conquer nature has at its ultimate aim the wish to conquer death or (to put it positively) to attain immortality. This means that we are witnessing in our time a sort of return to the view that immortality is somehow fundamental to us. As both Plato and Aristotle bring out, this desire naturally allies itself with both moral seriousness and religious devotion: so great a good as immortality surely has to be earned. And many serious people today contend that immortality is in fact available to everyone—to everyone, that is, who sees "the way and the truth and the light." But not a few would reply that modern natural science makes such belief impossible, it being allegedly the stuff of benighted fantasy. Yet if God is dead, the hope for immortality has evidently not suffered the same fate. That hope, however, is now more likely to be tethered to the entirely human activity of science.

I suspect that Plato and Aristotle would have several reservations about this. For example, no one can know in advance what other changes to human life the attempt to conquer natural death might bring about, and only the most naïve attachment to progress could dismiss such concerns from the outset. What is more, such a hope for immortality is, from the older point of view, exceedingly strange—for to conquer death on the basis of science is to wish to eradicate necessity on the basis of necessity. Let us be scientific! Let us rely on natural necessity—except the thought that "everything that comes into being necessarily perishes." In the *Theaetetus*, Plato quietly criticizes the flight from this world, even as he gives powerful voice to the longing that propels such flight, and he seems to counsel the resignation or acceptance that goes with self-knowledge. There appears to be today a newfangled version of this assimilation to God, which takes the form of our seeking God-like powers, and which is in its own way a flight from this world, every bit as much as is the religious one: the new scientific-religious zeal cannot accept this world as it is and must be—or human beings as we are and evidently must be.

Chapter 8

Xenophon and the Conquest of Nature

Christopher Nadon

One should admit from the start that the "conquest of nature" is not on the face of it a prominent or central concern in Xenophon's works. I believe Leo Strauss was the first to draw attention to this theme there. In his "Restatement on Xenophon's *Hiero*," an essay devoted to addressing the objection that the "classical orientation" has been "made obsolete by the triumph of the biblical orientation," Strauss claimed that present-day or modern tyranny is based on "the unlimited progress in the 'conquest of nature' which is made possible by the diffusion of philosophic or scientific knowledge."[1] According to Strauss, both possibilities—"the possibility of a science that issues in the conquest of nature and the possibility of the popularization of philosophy or science"—were known to the classics and rejected by the classics as "unnatural" in the sense of "destructive of humanity."[2] To support his contentions, Strauss asked his readers to compare a passage from the *Memorabilia*, in which Socrates appears to criticize those who seek to know the necessities responsible for each thing coming into being so as "to make winds, rains, seasons, and anything else of the sort they need whenever they wish," with a fragment from Empedocles that is likely Xenophon's source for the expression of this aspiration.[3]

There are several difficulties with Strauss's reliance on Xenophon here. First, if the conquest of nature is a distinctively modern project, then Xenophon would seem to be an unlikely opponent, especially as Strauss himself sometimes represents Xenophon as "the point of closest contact between premodern and modern political science."[4] Francis Bacon certainly found much to admire in the self-reliant author of the *Anabasis* who was willing to conflate "Zeus" with "chance." And Bacon's disciple Benjamin Franklin considered Xenophon's Socrates such a kindred spirit that he modeled his *Autobiography* in part on the *Memorabilia* and published rather free translations from that work for the edification of his fellows.[5] Poor Richard's claim that "God helps those who help themselves" found expression in Xenophon's pages long before it appeared in Algernon Sydney.[6] Second, while the passages from Xenophon and Empedocles may well show that "the possibility of a science that issues in the conquest of nature" was known to the classics, they certainly do not demonstrate that "the classics" rejected it as "unnatural," unless we are to assume that Xenophon was somehow more classical than Empedocles. These passages, along with Strauss's accompanying citation to Plato's *Theaetetus*, show at most that both the conquest of nature and the popularization of philosophy or science were debated and disputed issues in classical times. Third, and most interesting, it is by no means clear, even in the passage that Strauss quotes, that Xenophon's Socrates stands against "the conquest of nature." Instead, he raises it as a question whether those who sought "to know the necessities by which the divine things come to be" do so for the sake of controlling them, or did it suffice for them "to know only how each of these things comes to be,"[7] although he may well have separated out and opposed what often seems to us a necessary corollary or condition of the conquest of nature: the popularization of science.

The first chapter of the *Memorabilia* presents Socrates particularly in light of one of the charges that led to his death: not believing in the gods of the city but bringing in new gods (1.1.1; cf. 1.3.1). To understand the apologetic character of the defense against this charge, it is helpful to recall a story told by Herodotus, an author well-known to Xenophon.[8] When Harpagos was subjugating southern Asia in conjunction with Cyrus's rebellion against the Medes, the Dorian city of Cnidus decided to defend itself by turning the peninsula on which it was located into an island. But the workers digging the canal suffered a large number of injuries, especially to their eyes. The affliction appeared to be sent by the gods, so the Cnidians ordered a delegation to go to Delphi to ask the oracle why this was happening. The Pythia told them, "Neither raise a wall upon nor dig down through the isthmus; for Zeus would have given you an island if he had so wished." When the Cnidians received this response, Herodotus reports that "they stopped digging and gave themselves up without resistance to Harpagos."[9] He contrasts these

Greeks with the Babylonians who labored to make their city safe from the Persians by digging channels, creating an artificial lake, diverting the Euphrates, and making embankments along the river out of brick. Had they been accompanied by any vigilance, Herodotus thinks, these measures would have allowed the Babylonians to hold the city and destroy any Persians making the attempt "like fish caught in a trap."[10] If, then, at least for some Greeks, submission to barbarians was apparently preferable to altering the divinely arranged order of the world to ward off defeat, how much more impious to them must have appeared those who "conversed about the nature of all things" and examined "the cosmos" with a view "to making winds, rains, seasons, and anything else of the sort whenever they wish," and who in this way put themselves on the same level or perhaps even above the gods (*Mem.* 1.1.11, 1.1.15).[11]

By stressing how much Socrates differed from these kinds of "natural scientists" as a proof of his piety, Xenophon would seem to agree with popular opinion and all but concedes the impious character of their interest in seeking out the necessities (the material and efficient causes) of coming into being. In this, Xenophon's defense of Socrates differs from that of Plato. Whereas Plato's Socrates, at least in the *Apology,* himself brings up and assimilates to his own case the kinds of accusations "ready at hand and made against all who philosophize" so that they might also share in his ultimate vindication,[12] Xenophon is willing to leave the larger opinions of his fellow citizens unaltered. His concern appears more narrowly focused on the reputation of his own teacher; or perhaps he has less confidence in the possibility or even advisability of enlightening the Athenians. In any case, Xenophon also distinguishes Socrates from his predecessors both by the scope and method of his research: he conversed not about the heavenly but "always of the human things" and examined them by raising the question not of the necessities by which they come to be or pass away but of what they are, their class characteristics or forms; and he sought for answers by means of a dialectical investigation of what was said about them (*Mem.* 1.1.16, 4.6). Such an account allowed Cicero to claim that Socrates turned away from Anaxagoras and his efforts to measure the size and distances of the stars to become "the first to compel philosophy down from the heavens and set it in cities and homes and compel it to ask after life and morality and good and evil."[13]

Whether or not this turn away from natural science conclusively establishes Socrates's piety or belief in the gods of the city, there remains a more immediate problem. Xenophon's account here is incomplete as well as misleading, even if we limit ourselves to evidence taken from the *Memorabilia*. For in the penultimate chapter of that work, we learn that Socrates in fact encouraged even his gentlemanly companions to study geometry and astronomy to the extent necessary for properly apportioning and managing land (geometry), or for correctly marking time when navigating, hunting, or farming (astronomy). Given the close connection between hunting and war, and the fact that it is in war and farming that human beings are necessarily most exposed, Socrates knew that the study of these sciences was useful for the political community that depends upon both. He cites the Persian King Cyrus, the embodiment of political virtue in the Xenophontic corpus, as a model for a young Athenian gentleman to imitate precisely for his dedication to the arts of war and farming (*Memorabilia* 2.1.6; *Oeconomicus* 4.4, 5.14–15). Yet, especially as regards astronomy, Socrates limited himself to teaching only those parts of it that were directly useful; and he held that discovering "how the god contrives each of the heavenly things" was both impossible and "in no way pleasing to the gods" (*Mem.* 4.7.6). He advises one young gentleman to reflect on the fact that "the sun does not permit himself to be seen by human beings precisely, but takes away the sight of anyone who shamelessly attempts to behold him" (*Mem.* 4.3.14). Yet Xenophon also tells us that Socrates pursued these astronomical inquiries himself past the point of practical utility, even considering "the causes" (*aitias*) of the planets' motions and orbits.[14] Moreover, he shows that while Socrates disputed Anaxagoras's doctrine that fire and sun were the same thing, he did so not because such a claim was impious or displeasing to the gods but instead because it did not accord with more acute observations of the relevant facts (*Mem.* 4.7.1–7; cf. Plato, *Apology* 26d).

As Xenophon tells us in the previous chapter, "Socrates never ceased in the midst of his companions examining what each of the beings is" (*Mem.* 4.6.1). In other words, he did not in fact *always* limit his study to "the human things" but continued his interest in natural science. This conclusion is of course compatible with the earlier passage from Book I, where Xenophon claims that Socrates examined those who spoke about "the nature of all things" to discover whether they did so because "they thought it was fitting to neglect the human things or whether they thought they had already examined the human things sufficiently." Studying "the divine things," an inquiry that could "exhaust the life of a human being"—something which is never said of the study of the human things—might well then accompany or follow a completed study of the human things (*Mem.* 4.6.1; cf. 1.1.16, 12). But what, then, does Socrates learn from his approach to "the human things" that supplements natural science and perhaps explains why he limited himself from wholeheartedly teaching if not studying it?

Among the various "what is" questions that Socrates examined, Xenophon lists "what is a city" (*Mem.* 1.1.16). No chapter in the *Memorabilia* contains an explicit discussion devoted to this topic, but it is possible to infer something of Socrates's views indirectly (cf. *Mem.* 4.6.1). As with Aristotle after him, Xenophon's Socrates considers the city to be made up of a large number of households that are themselves dedicated to providing for their members' essential needs and enlarging their wealth (*Mem.* 3.6.14; cf. *Cyropaedia* 1.6). But unlike Aristotle, Socrates never maintains that the city is natural. The artificial or conventional character of the city, reflected in the various ways of organizing cities into different kinds of regimes (*Mem.* 1.2.41–46), stems from the household's and the city's initial dependence upon farming, which is itself an essentially collective undertaking (*Oec.* 5.14; cf. 15.4). While Socrates is willing to claim (and have others believe) that the earth, "who is a goddess," supports farming as "a soft pleasure"—one that brings forth easy nourishment for dogs and horses, teaches justice "by giving the most good things to those who serve her best," and calls forth the dedication of those willing to defend her when invaded by enemies—he also knows that this same earth does not provide easy nourishment for human beings but produces instead wild animals that ravage crops. And by "bringing forth and nourishing the necessary things outside the [city's] fortifications," where they are difficult to protect, she encourages a belief in the right of the stronger to take them, the very opposite of justice (*Oec.* 5.1, 5.4, 5.12, 6.5–7). Moreover, these crops are subject to "hail, frost, drought, violent rains, blights and often other things that wreck what has been nobly conceived and done" to such an extent that even the enjoyment of the pleasure gardens (*paradeisoi*) of the Persian king can be disrupted by irregularities in the seasons, since "the god does not manage the year in an orderly way" (*Oec.* 4.13, 5.18, 17.4). And should rain happen to fall at the right time, "weeds come up with the grain and choke it." If not done properly, efforts to strengthen the land by purifying the fallow result in even more weeds; and the toil necessary to engage in this endless struggle comes from slaves who must be controlled both by arms and forensic rhetoric (*Oec.* 5.13, 16.14, 17.15, 20.20).

For a city to survive, it must therefore be organized and structured in such a manner as to control the land or nature, to compel it to produce by means of art (*technē*) what it would not otherwise bring forth. The city must regard and treat nature as do the natural scientists Socrates apparently criticized for their desire to master nature through a grasp of material and efficient causes. Yet Socrates also teaches his fellow citizens to think of the land as "a philanthropic goddess" who teaches a form of justice (the generous reward of good service) that she does not herself always practice. As such, she must be appeased by sacrifices offered in the spirit of submission. The city must act as though the earth were an enemy while also treating her, at least in speech, like a friend or benefactor. Ischomachus, the Athenian from whom Socrates learns what a gentleman is, embodies this tension or conflict. On the one hand, he takes inspiration from and imitates the Phoenician boatswain who deploys human art and foresight to protect his ship "when the god raises a storm at sea," and, like his father, he works to improve the land. On the other, he expresses the view that "there is no advantage in fighting against the god" whom he also considers responsible for "the nature of the earth" (*Oec.* 8.11–17, 17.10, 20.22–25, 16.2–3). But if the political community is constituted on the basis of this tension, then it depends not only on technical arts, like agriculture and metallurgy, but on the art of rhetoric as well; and this rhetoric is not merely or even principally for the consumption of the unwilling, hard-worked slaves.

What is it that keeps the city from abandoning its respectful attitude to a divinely ordered nature it must otherwise exploit? And why is it that Socrates is willing to leave and even contribute to his fellow citizens living in such a state of mind rather than enlightening them as to their true situation? At first glance, Socrates seems to teach a continued reliance on and deference to the gods because the arts themselves are far from perfect or self-sufficient. Farming is always dependent on the land and even the best, most diligent farmer can have his crops fail due to circumstances beyond the control of his art and care. This impotence leaves room for divination and propitiation of the gods through sacrifice (*Oec.* 5.19–20; *Mem.* 1.1.8–9). But not all arts are equally connected to the land and hence equally subject to its vagaries. Socrates knows that those practicing the so-called banausic or mechanical arts, and who lack the farmers' bond to the earth, "who is a goddess," likewise, due to their independence, lack the desire to toil and run risks to defend her (*Oec.* 6.5–7; *Hellenica* 7.3.12.). Might there somehow come to be a city of such artisans who could then be taught a proper appreciation, or deprecation, of nature? Perhaps not, given the extent of dependence of all other arts on farming: "Whoever said that farming is the mother and nurse of all the other arts spoke nobly indeed. For when farming goes well, all the other arts also flourish, but when the earth is compelled to lie barren, the other arts almost cease to exist, at sea as well as on earth" (*Oec.* 5.17). Still, might not the art of farming itself come to be perfected over time so as to achieve the kind of autonomy or control typical of the mechanical arts? And would the time it takes to reach this end not be greatly diminished if farmers were taught to have a less reverential attitude toward the land and a more appreciative one of the natural scientists who might help them? Notwithstanding the admiration shown for them by the author of *Poor Richard's Almanack*, Xenophon and his Socrates seem to stand in the way of such a solution.

Xenophon's best regime is not a city of artisans but of soldiers and farmers, an improved or idealized version of Sparta transplanted to some far away barbarian place: Persia. Its citizens pride themselves on having a political regime that produces the best men through a system of education that cultivates the virtues and includes a school of justice, not science and engineering. Those who make it through the various levels of education are considered perfected or completed men. By law, the schools are open to all. In practice, only those families that can somehow raise their children without putting them to work can send them. The result is that very few finish the education, and these numbers are made even smaller by the fact that Persia is a country with very poor land. On occasion, when most generous, it gives back but a two-fold increase in grain (*Cyro.* 8.3.38). The inability of the Persians to provide sufficient leisure for education not only limits the number of men who can be made good, it also determines to a large extent what the Persians consider to be a good man. A meritocratic aristocracy in principle but an oligarchy of families in fact, Persia is ruled by men who live off the forced labor of subsistence farmers. The presence of so many disenfranchised compels the Persians to cultivate and practice those qualities required to keep them under control. They call these virtues. For example, compelled to camp out around the buildings that contain the city's arms, they make a virtue of this political necessity and call it "moderation." Even the justice they practice is more for the sake of preserving inherited wealth than for giving to each what he can use (*Cyro.* 1.3.16–17; *Oec.* 1.14). But justice understood as obedience to law, while essential to the political stability of the city, is never optimal for the individuals who comprise it. And the practice of this kind of political virtue can prove harmful to the one who has it, even when he is compelled to acknowledge his gratitude to the fatherland (*Cyro.* 1.2.7, 4.1.8; *Mem.* 1.1.8, 2.6.39, 3.3.10, 3.7). For this reason, the Persians teach their children, much as Socrates does his companions, to care for the gods and, after having made themselves such as they ought to be, to trust that they will be propitious and ensure that dedication to the city will turn out well (*Cyro.* 1.6.3–6; *Mem.* 1.1.9, 2.1.28). The political community therefore depends on something above or at least outside itself to guarantee or support a belief that the life it requires its citizens to lead is actually good.

Xenophon's Cyrus takes advantage of this gap between the necessity and the goodness of political virtue to convince the Persian elites to abandon virtue, understood as making them complete or good men, and to practice it instead simply for the good things he promises can come from it: "much wealth, much happiness, and great honor for themselves and their city" (*Cyro.* 1.5.9). But at the beginning of his project, he refrains from disclosing to the Persians that his own ambition in fact already transcends their city, which will be irreparably destroyed by the universal meritocracy he attempts to establish without respect to ties of citizenship, fatherland, family, or tribe (*Cyro.* 1.6.9, 2.2.26). Cyrus seeks not to conquer nature but to conquer the world. But to do so, he must conquer or attempt to change what might be called the nature of the city or political life. What Shakespeare's Coriolanus does unwittingly, Cyrus consciously accomplishes when he too "unroofs" the political community in an attempt to overcome its internal contradictions and reliance on external powers by establishing himself as a providential king, one who might perhaps even be thought to rule on the basis of science or knowledge (*epistamenōs*).[15] But Cyrus's empire proves in Xenophon's treatment to be a spectacular failure (*Cyro.* 8.8). Just as the Persian republic was unable to establish a genuine common good between rulers and ruled, or even among the rulers, so too does Cyrus fail to establish a common good between himself and his subjects, and necessarily so given the ends they pursue. Cyrus cannot make what is good for him also good for all his subjects, any more than the laws in Persia could make what was good for the city always good for its citizens. If this contradiction is characteristic of all political communities, then they cannot be ruled simply by reason or knowledge. There can be no political science in the sense of a scientific art of justice or even of ruling. Nor can political virtue be taught simply by speech or reason (*logos*).[16] But if all political communities are self-contradictory, they are not all equally unstable. Cyrus's empire is especially unstable because the conflict between the good of the ruler and the ruled, unmediated by the rule of law, is particularly manifest (*Cyro.* 8.4.31). Moreover, this tension is aggravated by what Cyrus teaches about the gods, whether as his genuine opinion or perhaps as a means to justify his own policies: "Not even I am able to overcome the insatiable desire for wealth through which the gods, by putting it into our souls, have made all equally poor . . . thus I serve [*hupēretō*; alternatively, "submit to"] the gods by always grasping after more (*Cyro.* 8.2.20–21; cf. 4.1.14). Impiety that leads to lawlessness characterizes "everyone in Asia" living in the wake of Cyrus's empire and example (*Cyro.* 8.8.5).

Socrates, however, holds and teaches that "to need nothing is divine, that to need as little as possible is nearest the divine, and that what is divine is best, and what is nearest to the divine is nearest to what is best" (*Mem.* 1.6.10). He does this not because he has a more certain or direct access to understanding the true nature of the divine or of the gods than Cyrus, but rather because his study of the human things has taught him that the city requires something outside or beyond it, both to support and limit what it necessarily must claim to be human virtue; and his teaching about the divine is in greater harmony with this need than is the unlimited and destabilizing views about the gods, as taught both in speech and in deed by Cyrus. This political criticism of Cyrus's "theology" applies likewise to those pre-Socratic natural scientists like Empedocles, whose desire to understand and control the divine things through understanding the necessities by which each thing comes to be both is and appears to the city to be, at bottom, atheistic and hence also corrosive of political life.[17]

Xenophon depicts Socrates consistently acting on this insight into the city's dependence throughout the *Memorabilia*. Despite, or perhaps because of, his open and extreme criticism of fathers (*Mem.* 1.2.52–55), he reminds both of the philosophers he converses with in that book (Aristippus and Hippias) of the need for them to understand the city not simply as a regime existing by convention and subject to manipulation and change but also as "a fatherland" that demands respect and calls forth devotion as something standing above human making or control, and therefore also provides some ground for "great hopes" (*Mem.* 2.1.19, 4.4.14). While Socrates makes no mention of or demands any respect due to himself, he does advise his son not to show ingratitude to his mother, Xanthippe, for fear of being punished by the city (*Mem.* 2.2.13). A companion of Socrates, skeptical to the point of openly mocking the city's religious practices, is subjected to a semi-public lecture on intentional design and providential gods that Socrates himself does not believe, given its internal contradictions (*Mem.* 1.4); and based on his own observations, Socrates knows that "the god fails to manage the year in an orderly fashion" (*Oec.* 17.4). Accordingly, he also encourages those useful arts that provide or hedge against this neglect. But there are better and worse ways of doing so.

Cyrus, too, seems to recognize there are problems with the weather. He almost completely remedies the defects of the seasons by conquering an empire so vast that he can winter in Babylon, move on to Susa as the temperatures climb, and then escape the peak of summer heat in Ecbatana on the edge of the Zagros Mountains: "Acting in this manner, they say that he spent his time in the warmth and coolness of a perpetual spring" (*Cyro.* 8.6.22; *Oec.* 4.13). Yet an empire that allows and supports such movement proves unsustainable, not least of all because such pleasure gardens necessarily provoke envy and the measures necessary to cope with it (*Cyro.* 7.5.76–77; cf. *Mem.* 1.6.3). Socrates counsels a less perfect but more stable remedy against the imperfect ordering of the year when he advocates building houses with windows and eaves designed to exploit the sun's rays in winter while guarding against them in the summer to make the house pleasant in all seasons. If Socrates's cost-benefit approach to architecture threatens to reduce "the noble/beautiful" (*to kalon*) to the useful so as to eliminate "paintings and embroideries" (the noble/beautiful) altogether from private dwellings, he is careful to mitigate this tendency by also recommending that temples and altars be built in visible (public) but out of the way places, since "it is pleasant for those seeing them to offer prayers, and for those undefiled to approach them" (*Mem.* 3.8.10). In this, Socrates consciously imitates and improves upon the unconscious and imperfect practice of the gentleman-farmer Ischomachus, who tries to reconcile the endless struggle against the weeds put forth by the earth ("who is a goddess") with his counsel not to struggle against the god, by characterizing this labor as a "purification of the earth," as if man and not the god were somehow responsible for its polluted condition (*Oec.* 16.3, 16.13, 20.13, 20.20).[18]

The city must both exploit and respect nature. It must both encourage and restrain industry, the human capacity to take matters into our own hands and care for ourselves. In keeping with this classical, or rather Socratic, insight into the necessarily contradictory and imperfect character of political life, Strauss misled his twentieth-century readers into believing that the classics, or at least Xenophon, simply opposed the conquest of nature as destructive of our humanity. In our day, moderation perhaps calls for greater restrictions on industry. But this is a political judgment. Yet there being no one, eternal, unchanging law to govern or guide us, it is possible that, in other times and circumstances, where men so distrust their faculties as to sap all human initiative, this same insight could lead one to counsel and encourage such a conquest. More worrisome to Strauss and Xenophon's Socrates is the "popularization of philosophy or science" (Enlightenment) that makes political life so much more difficult to manage if not rule. Xenophon, perhaps more than any other philosopher of antiquity, guarded against this danger. Strauss may well be his closest modern imitator.

Chapter 9

Lucretius on Rebelling Against the "Laws" of Nature

Paul Ludwig

Lucretius seems closer to modern science than most other ancient philosophers. Yet he does not believe in mastering nature. Why not? For Lucretius, we humans rebel against nature—against our natural propensity to die—creating thereby many unnecessary pains for ourselves, disturbances of mind, including especially the disturbance of fears and unnatural hopes surrounding religion: hopes for a continuation after death, hopes and fears that our justice will be rewarded and our injustice punished. Today, the religious rebellion continues. But, to religion, a new rebellion has been added: a scientific rebellion. The scientific rebellion against nature—an attempt, in part, to defeat death—bases itself on a view of nature very like Lucretius's and, in an irony of history, owing something to his influence. The modern project has requisitioned his atomic view of nature, by and large, without taking the ethical consequences—the way to live happily—that came with it. If we could defeat death, our happiness in this life might not depend on coming to grips with the reality of death and accepting it, nor on achieving the Epicurean goal: tranquility of mind.

Laws of Nature

I say rebellion because Lucretius is an early proponent of natural law. Broadly speaking, classical Greek philosophy, dominated by Plato and Aristotle, did not recognize "laws" of nature. The very term is an oxymoron: law is identical to convention. Nature is the opposite of convention. Fire burns the same way in Greece and in Persia, but Persian conventions are very different from Greek ones. In Aristotle's natural philosophy, the substitute for law is form. Classification by forms—what geology still does a lot of today, incurring mild pity from more advanced sciences—sorting entities into classes (granite is an igneous rock; sandstone is a sedimentary rock; etc.), is much of what ancient natural philosophy consisted of: getting the essence of the thing right, saying what it is. Tigers and leopards are both cats. But a leopard is not a tiger, so each is a species of the larger genus. Species is a Latin word for form. Lucretius speaks several times of *ratio*, reason, in tandem with *species*, form.[1] For Plato and Aristotle, the form *is* the reason of a thing. For Lucretius, the species are not eternal, and he speaks of the "reason *and* form of nature," raising the question whether there is one uniform account and form of nature or whether separate kinds have natures.

For Lucretius, atoms are eternal. This is the reality: atoms fall, straight down, in a void. The void is infinite, so the fall is eternal. There are an infinite number of atoms. Every now and then, inexplicably, an atom swerves in its fall and collides with another atom. Since one atom's colliding with another sometimes produces a compound, it follows that some atoms must have hooks that can snag or catch on other atoms' hooks. Such compounds explain how the material world we perceive gets formed. Lucretius compares atoms to the motes of dust that we can sometimes see when sunlight comes shining through a window at the proper angle. The motes of dust are flying around randomly, but, just by accident, every now and then, a dust mote collides with, and sticks to, another dust mote (2.114–124). That is the creation of the cosmos. It has already happened an infinite number of times previously, and it will happen—both in this same combination and in an infinite number of other combinations—again in the future, an infinite number of times. The destruction of the cosmos is also inevitable—a coming apart of what once came together. Are there, then, laws of compound? Some atoms can compound with others, and others cannot compound with each other. In theory, one could figure out which ones compound with which and thereby create a table of elements. Lucretius is interested in what can compound with what but far *less* interested than early modern chemists, many of whom had a practical aim: to transmute baser metals into gold.

Lucretius's laws of compounding are not truly atomic, and they have two or three practical purposes. The first (1) saves him from the fanciful, chimerical compounding that religion alleges. Centaurs, for example, are not possible (2.700–717; 5.878–924). Knowing nature well enough to prevent religious speculation is one of his main goals. Perhaps a subcategory of (1) is a second purpose: (2) knowing the laws of compound well enough to know that the soul is mortal. And finally, one could argue that a third goal of knowing rules of compound is (3) knowing what is truly a pleasure and what is not (2.20–61, 5.1412–1435). Such uses of compound suggest a difference with early moderns, and the advantage would seem to go to Lucretius. From Lucretius's standpoint, early modern chemists or alchemists had not philosophized about themselves and their own souls sufficiently to cure them of the desire for money. A philosophic soul does not desire much beyond meeting subsistence needs and a few simple pleasures. The pursuit of wealth cannot make you happy.

It would be remarkable that Lucretius speaks of natural "laws" if a law presupposed a Lawgiver. Lucretius is more opposed to this divine origin theory than to practically any other single thought. Looking up at the heavens and thinking that it must be the design of a Designer, instead of coming from chance, is the mental move that causes us to believe in gods and, thereby, to ruin our lives, giving ourselves "harsh taskmasters" (6.58–67). There was no plan that those atoms ever swerved to hit one another. Appearances are against his view: species, like wolves and violets, look substantial and purposive. But that is random; such apparent design could *only* have come into existence by chance, according to Lucretius. It is certainly great good luck that our universe is as intelligible as it is, but Lucretius deals with this objection by infinitely increasing the chances of its coming into being through his appeal to the eternity of time and the inexhaustibility of material (5.187–194). Every combination has been tried; we just happen to live in one that looks more successful than the vast, infinite numbers of more barren combinations. Barrenness exists in our world, too, which contains far too many defects for it to have been created by benevolent gods.

Lucretius's preferred word for natural law is not *lex* (nor *ius*) but *foedus*: literally, a compact or covenant. It is through *foedera*, treaties or pacts, that the things we see around us come about—things which his poem "On the Nature of *Things*" claims to explain, or to give a reason (nature) for. For example, in his famous theory of the lodestone, he says, "I will begin to go through, by what covenant of nature it comes about that this stone can draw iron, this stone which Greeks call Magnet from the name of its fatherland" (6.906–908). In more general terms, he claims to teach "by what covenant all things are created [and] in which it is necessary for them to remain and [why] they are unable to rescind the strong laws [*leges*] of time" (5.56–58). I would speculate that, although a law might presuppose a Lawgiver, a covenant is democratic; it presupposes no higher power than the participants themselves, no king. So covenant, convention, is preferable because it offers no foothold for a Creator, which could potentially lead into religion.

The participants in the covenant are atoms. Lest we warmly welcome this sentiment about participatory democracy—deliberative democracy—he later warns, "For certainly neither by a plan did the primordial things, all of them, with a sagacious mind, locate themselves in their order, nor did they all make a bargain about which motions each should produce" (5.419–421). Instead, atoms convene to make their covenants in a kind of warfare of clashes. In fact, some of the major compounds that earlier thinkers had mistaken for elements—fire, water, air, earth—have in the past won out or overcome their opponents and temporarily overwhelmed our world, or parts of it, for example in myths of the flood, or in conflagrations (5.341–342, 385–395). One or other of such compounds may well win out in a more final way in the future, at the end of the world. The balance they have achieved is a temporary truce, a pact, an uneasy agreement to let things remain *status quo* as each gathers its powers for a future, final confrontation.

Covenants of nature have the drawback of neglecting a difference between human covenants and natural ones: the element of mind or reasoning, the ability to deliberate, which humans appear to have and atoms explicitly do not. What does Lucretius think of mind? He uses *foedus*, *ius*, and *lex* to denote human conventions, too, human constructs, in his speculative prehistory of mankind. There was a time when humans did not respect or see the "common good" (his statement implies they see it now, or can do so, at the time of his writing) and when humans knew no *mores* nor how to use laws (*legibus*; 5.958–961). Eventually (5.1025), men kept covenants (*foedera*), which they made in primitive language, covenants whereby they did not want to harm or be harmed, thinking especially of the safety of their wives and children. *Laws* came in much later than covenants. An intervening period of kings, by which he must mean unconstitutional monarchs, kings without law, came before law. Finally, during a period after the kings were slain (1136–1144), at the end of a war of each against all, "some" taught men to create magistrates, and these same teachers "constituted *iura* in order that men might be willing to use *legibus*." So covenants are prior to laws, both in nature and in society. They are chronologically and perhaps logically prior, in society. In nature, they are logically or lexically prior or preferable—but again, for the purpose that we do not backslide into theology.

From Lucretius's point of view, modern science's natural laws might show a dangerous religious tendency. Of course, in modern science, the use of the term law is an *analogy*. But why this particular analogy? Why think that natural phenomena must obey certain rules—or else, what, get punished? The divine origin is presupposed in medieval Scholasticism; it is appears to be presupposed in early modern "Book of Nature" teachings, such as in Galileo's famous sentence. A book requires a Writer. Again, however, modern scientists and historians of science would argue that we are no longer beholden to the religious origins of our jargon; if we make the proper distinctions, perhaps there need be no religious contamination.[2]

Saving the Appearances? Atomic Ethics

What is the status of these temporary atomic covenants? Are they "real"? Covenants are at least "things" (*rēs*). Yet things seem to lack reality, if atoms are the ultimate reality (nature). The things that appear to us to be at rest are really in motion. Lucretius compares such apparent rest or wholeness to a herd of sheep seen at a distance, and to an army on maneuvers, playing war games (2.308–332). A herd or an army, too, *looks* at rest, from a distance. Entities that appear stable are like the fluffy white mass of a herd on a hillside, which is imperceptibly composed of individual members moving about randomly. Nothing is at rest. Atoms are the individuals in this analogy: each sheep (or soldier) is imperceptibly a mass of atoms moving randomly. Where does that leave their reality, the reality of compound bodies and of species?

Could only atoms have natures: are they the only natural things, if compounds are merely covenants and, hence, conventional?[3] Lucretius seems to use the word nature in more than one sense: there is the nature of "the things" as well as another hidden and unseen nature. Only this hidden, atomic nature preserves the nature of the things *qua* their being perishable: the imperishable proves that all else is perishable. At 1.684–687, the nature of fire changes (into something else) when the bodies underlying fire change. In arguing against the doctrine of four elements, Lucretius says the four would not mix adequately, and you would *see* bits of each of the four, separate, in any given thing, but this does not happen. "But the *primordia*, in producing things, need to apply a nature secret and unseen [or "clandestine and blind"] in order that nothing may stand out which fights against and prevents each thing from becoming what it properly is" (1.775–781). Anaxagoras erred in this regard, supposing *primordia* "endowed with a nature similar to that of the things themselves" (1.847–849). Anaxagoras thought that some parts, like flesh, were flesh all the way down—if you keep subdividing, you do not get down to new particles but you just keep getting smaller and smaller bits of the same stuff. For Lucretius, the primary nature has to be different. Only that difference secures that some nature is unchanging. Only the underlying matter can be eternal.[4]

The two natures raise again the question of the status or reality of the visible, changeable nature. Species, including the human species, are sufficiently stable (Lucretius's example is kinds of birds with their different markings) to justify the description that "*nothing* changes, but all things are constant *to such a degree*" that "a limit" has been given for growing and holding onto life: "It stands having been rendered sacrosanct, through conventions of nature [*per foedera naturai*], what each thing can do and what it cannot" (1.584–590, emphases added).[5] We note the religious rhetoric.

Lucretian science, like modern science, fails to save certain appearances. Lucretius does not explain satisfactorily the emergence of living beings from non-living beings. Nor does he account for the next great boundary, that between irrational life and rational life. Lucretius is perhaps more thoroughgoing than his modern counterparts on such questions. For example, Lucretius did not believe that plants are alive.[6] Lucretius did not believe—of the human mind and free will—that they differ in kind from horses' minds and free will (2.263–268).[7] The failure to preserve these distinctions has ethical consequences. For example, another atomist, Democritus, saw no reason for children to be grateful to parents (because nature compels parents to love them: why should you reward someone for doing what they are forced to do?[8]). Nature destroys conventional ethics. The deterministic thinking of modern science, at least, has its normal ethical result in people's living down to new, lowered expectations. Instead of being bound within Tocqueville's "fatal circle"[9]—the exact limits of which they are unsure of, and within which they are powerful and free—people instead bind themselves into a narrower circle, one created by their own ideas. Who can hold himself to standards if he believes himself nothing more than an assemblage of "selfish genes"? The dissolution of appearances leaves us with poor choices: we can either reduce life and rationality to something less than they appear to be, as Lucretius does, or we can elevate inanimate beings into sentient and rational life, like Leibniz did when he considered even stones to have a low buzz of sentience (contrast 1.915–920). Are we forced to choose, if we think deeply?

Our status or reality as compound bodies, and as a species, is thus bound up with the question of how we should live, how to be happy. Should people imitate the atoms, colliding with each other blindly? Or should we master the unruly atoms, impose our own ideas on them? But where do those ideas come from? As Strauss ("Notes on Lucretius" 94) writes, "The first things are in no way a model for man." Is there any sense in which the Epicurean philosopher is the most natural of men or lives most naturally—not in rebellion against nature? Lucretius suggests that we focus on simple pleasures, pleasures mostly of the body, learning what the limits of pleasures are. He recommends a kind of ascetic moderation. Any pleasure that perturbs the mind, unbalances the soul, is not truly pleasant at all. For example, romantic love, by mixing pleasures with pains, leads at best to a very impure pleasure.[10] Lucretius describes one prominent mental pleasure, of which he says "nothing is sweeter": pleasure in the doctrine of the wise when, through it, one feels safe and unperturbed oneself while viewing others who have gotten themselves into dangers and perturbations of mind through their not knowing which pleasures to seek; he compares it to a man on shore looking out to sea at others' troubles (e.g., on sailors desperately trying to bring their ship safely to shore), or looking at a battle raging (2.1–19). He especially has in mind ambition for riches and a life spent jockeying for political power. "Nature barks"—rages or roars, gapes (*latrare*)—"for nothing other than this for itself: that pain be absent from the body, and the mind reap an agreeable feeling, when removed from care and fear." He must refer here to the "seen" nature, the "thing" that is us. The serenity he recommends is obviously *not* to act like atoms—the soldiers and politicians in conflict with one another are acting more like atoms. Most attempts at mastering our circumstances and changing them would mean engaging in such conflict. He is certainly in favor of self-mastery, changing one's own soul, and probably, if he follows Epicurus with his famous garden, Lucretius would be in favor of arranging a private life for oneself and one's friends in which to live together and philosophize.

The Theology of Physics

Like early moderns, Lucretius presents himself as a benefactor.[11] In particular, he presents Epicurus as the world's greatest benefactor, especially for freeing mankind from religion (e.g., 1.62–79, 3.1–25). "Nothing comes divinely from nothing" is Lucretius's opening maxim, the gateway to his philosophy. If we remove "divinely [*divinitus*]," it is a close translation of Epicurus.[12] Does the maxim have any other content than to eliminate the possibility of divine action? Lucretius himself flirts, at least, with denying his principal maxim, seriously considering whether the moon, for example, in disappearing and reappearing, might actually be destroyed and reformed each time (5.731–750)—*any* such farfetched explanation is preferable, he seems to imply, so long as it is natural, and there is no recourse to gods, no coming about "divinely." He also says it is better "to reason lyingly" or "to give causes falsely" (4.503) than to abandon empiricism. From a modern perspective, which demands proofs, such a statement resembles the plaintive assertion of Piggy in *Lord of the Flies*, when faced with the murderous behavior of the other boys stemming from their superstitious dread: "Life . . . is scientific." Piggy, and Lucretius, cannot say how the world is scientific, but they take it as a certainty that there must be an explanation we do not yet know, a naturalistic account that explains any apparently supernatural occurrences.[13]

Yet if Lucretius posits things about nature in order to vindicate a preferred ethical result (freedom from religion), then modern scientists or historians of science must inevitably think his natural science is the very opposite of science. The logical order of proof becomes suspect.[14] Plato and Aristotle often stand accused of a comparable prejudice, merely with different ethics governing. Placing ethics highest is a way of ensuring that philosophical researches have meaning for our lives. Epicureans are not interested in science for its own sake.[15] Modern scientists have achieved near-indifference, though not complete indifference, to the religious question that so exercised Lucretius. His natural philosophy provides the basis for rejecting religion, in the rhetoric of his poem. But the actual rejection precedes the natural philosophy, which presupposes it.

If miracles are permitted, he argues, if some things could come from nothing, then anything could—and would. We would see miracles everywhere: "For if things came to be from nothing, every kind *could* be born from all things, nothing *would* need a seed. In the first place, men *could* arise from the sea, the scaly kind could arise from earth . . . ; cattle for plowing and other herd animals and every kind of wild beast, with uncertain birth, *would* hold cultivated and deserted lands" (1.159–164, emphasis added).[16] We note the slide from "could" to "would" in his argument. It is easy, however, to propose an alternative account, a middle way in which miracles occasionally change nature, affecting science in such a way that it is limited but not destroyed. On such a view, no "theory of everything" would be possible, in today's language—no master science with the ability to build up from basic axioms to predictions of everything that has happened or ever will happen in the world. Such a totalizing science would be impossible because, even if one great causal chain could be found, nevertheless the occasional miracles, posited in this alternative, would each have started new causal chains that would distort the overall chain. Science could continue to predict, for example, the local motions of a satellite in space; scientists monitoring it would merely have to hope that no particular miracle intervened during their observations. Lucretius's argument overstates the unintelligibility of a world in which miracles occur. His science proves that miracles do not always occur. It would require more premises to prove they never occur.

Perhaps because of such difficulties, Lucretius theologizes. His theologizing stands in stark contrast to the insulated, anesthetized approach of modern science, where God-talk is banished on methodological grounds—even in the field of cosmology, that study of the whole which would most naturally lead to a choice, a decision about theology. Science today leads many to doubt the existence of God, but such cause and effect may invert the original order of the rebellion and only partake, as it were, of the rebellion's rhetoric. Lucretius for his part pretends to continue to believe in gods so long as his readers are not so impious as to attribute to the gods any care for man or any motive to intervene in human affairs (1.44–49, 2.646–660, 3.18–24, 5.146–186). For such divine intervention would show that gods have needs, which would in turn contradict the perfection that piety attributes to gods. Lucretius, in other words, does not base his every assertion about the gods on atomic science. He begins instead, in this crucial case, from opinions commonly received, and merely refines these in accordance with what else people are saying.[17] The fundamental thing said about the gods, or the fundamental thing theologized, is the gods' perfection. If all activity presupposes need, then anyone who wants interventions, miracles, must give up perfection. The implicit premise concerns generosity or divine beneficence: generosity is never quite what it claims to be, a pure gift with no strings attached, but is always a disguised self-assertion or fulfillment of need.[18] In St. Thomas Aquinas's terms, Lucretius would deny that greater perfection is evinced if a being spreads its goodness around than if it enjoys its own goodness exclusively.[19] Via this premise, a naturalistic claim or account does seem to have entered the theological argument, which is not, to repeat, based on natural science. It is unclear how much this naturalistic premise owes to atomism. Its opposite, the sociality of perfection, seems to come from a human model. The Lucretian gods, too, seem modeled on the serene indifference of earthly, human Epicurean philosophers. We must consider, then, whether an original revolt against the gods is foundational to Epicurean philosophy rather than being within the scope of the zeteticism, the searching after truth, which that life entails. The foundational assumption might govern the project of inquiry while the inquiry itself can at best deepen and strengthen, but not finally vindicate, the foundation.

Lucretius and Modern Mastery of Nature

In considering what light Lucretius can shed on the modern project to master nature, I am left with more questions than answers.

1) The eternality of matter, presupposed by Lucretius, seems to have been superseded by modern science. In place of matter, "forms"—formal properties of laws—now seem to take precedence. Even natural laws seem to change, over time: laws for bigger, slower objects seem to have arisen in time from laws for small, fast objects, that is, from quantum mechanics. Does or should modern physics continue to consider itself materialist? Why does this matter? There is nothing eternal anymore. If nature is not matter, what is nature? The powerful *temporal* aspect of modern physics and evolutionary biology lends support to the way history has supplanted philosophy in modern times—and not only in the historicism of the major philosophers themselves. Has the very idea of nature changed?

2) Would this new nature—temporal rather than eternal—leave a stronger foothold for a divine Creator than Lucretius and classical philosophy generally believed? If things do come from nothing, and fundamental particles continue to come into being from nothing and go back into nothingness, where does that leave the principal maxim ("nothing comes divinely from nothing")? Initial resistance to the Big Bang was based, I am told, on concern that it seemed to imply an initiator or Creator. Such a concern would be Lucretian. Physics no longer requires the infinity of chances needed to produce intelligible order. Only fourteen billion years were required, and not all combinations were tried before this order came about. In each of these ways, God becomes more of an option, for scientists, a "personal choice" that also plays into our democratic freedom—different strokes

for different folks. Does the new science make the choice between religion and philosophy no longer as stark as it once was, thus changing the character of philosophy?

3) On the problem of materialism and life: Aristotle thought or said that living bodies were entelechies; each was a stasis of a sort, but a vibrant stasis—a working to stay itself. Precisely because the elements tend to war and move apart, a living entity must be "at work," "holding its aim or goal in itself" (*Metaphysics* 9.1050a21–23). Each thing is, in the end, nothing but that working to maintain its form. Aristotle's account seems better than Lucretius's in that living beings appear to *use* their atoms while not being identical to them. Hans Jonas cites the famous "ship of Theseus" problem, in which the Athenian state ship plies back and forth annually between the city and the Delian sanctuary, getting a plank replaced here and a mast there until finally, after many years, the whole ship has entirely new timber, new matter. Yet we *say*, at least, that it is the same ship.[20] The difference between ships and animals is that living beings do this replacement (of matter) all by themselves. But then, must we not say their activity—the activity of replacement, the activity that maintains the form—is the best description of their lives, not the material?

4) Does the therapy or therapeutic nature of atomism, both ancient and modern, make it less than philosophic? In modern terms, Lucretius's science is applied, not pure—applied to the goal of tranquility, equanimity, rather than being for its own sake. Epicurus in a letter said that prudence is even better than philosophy because prudence is necessary for all the virtues.[21] He also said, "If our dread of the phenomena above us, our fear lest death concern us, and our inability to discern the limits of pains and desires were not vexatious to us, *we would have no need of the natural sciences*."[22] Is this a class (species) trait of ancient and modern atomists: to think there is something better, out there, than philosophy? Modern science had its beginning in a turn to practice, away from contemplation (though the ultimate aim of its founders may have been to rescue contemplation). The practical application of science has also led, however, to great interest in pure, not applied, science; many scientists today are contemplatives, in a sense, far more interested in discovery of truth and in the beauty of discoveries than in using those discoveries for any purposes of society. Practice, then, was conducive to theory. From Lucretius's therapeutic point of view, such pure scientists fail to apply their knowledge to their own ethical lives, fail to find what will make them truly happy. A philosophic substitute for immortal fame and for making

one's name live on through progeny was once thought to be contemplation, in which the philosopher temporarily gets in touch with the eternal.[23] Contemplation itself can be therapeutic. But if nothing is eternal, then the relation between contemplation and therapy changes.

5) Does atomism *permit* self-mastery and a tranquil confrontation with nature? That is, if all things are truly "matter in motion," then mental life will be in motion, too; reason will be the slave of the passions; desire will follow desire; and appetites will be insatiable—making equanimity impossible. Not only citizens will be enslaved to the passions, but the philosopher will, too. In examining the bridge between Lucretius and modernity, Paul Rahe puts his finger on this fundamental difficulty in Epicureanism, a contradiction he thinks Machiavelli corrects.[24] Philosophy conceived as therapeutic control of passions will be impossible. Would not the modern project, then, be fundamentally superior to ancient philosophy—because ancient philosophy lacked consistency?

Conclusion

Descending now from questions to certainties: what can we take away, today, from Lucretius's teaching? One controversial aspect of the modern project is the attempt to defeat death. Let us suppose medicine will defeat death, and we could begin living indefinitely long lives. Still, when our sun begins to run down, and the solar system it created begins to crumble, we will run up against the problem of death again. Modern science agrees with Lucretius on this last point—whether it be the entropy of thermodynamics or the expanding universe that grows ever colder and winds down, or the destruction of everything in a Big Crunch. And yet, from the same information he had, some among us insist on drawing a different, less defensible conclusion than Lucretius's. If life cannot, even in theory, be infinitely prolonged, why not come to grips with death now? It might even be altruistic to do so: by living on indefinitely, we take up space and resources that must, in effect, be denied to future generations; we smother or suffocate future generations by not letting them live. Lucretius brings this argument down to the atomic level (3.964–971). Now, just to be clear: altruism is *not* Lucretius's main or even a very important reason for accepting death. Lucretius, true to at least a piece of what the popular epithet "Epicurean" suggests, is recommending that each of us secure his own personal happiness. Struggling against death is not a recipe for securing happiness. For it necessarily opens the door to infinite desires, for unnecessary things, a self-imposed Hell, Lucretius says, likening it to giving ourselves the punishment of Sisyphus, rolling a rock up a slope only to be thwarted and to have to do it all over again, endlessly.[25] Such is the pursuit of happiness. Unless we address the "leaky vessel" that is our mind, we can never hope to be fulfilled (6.17–21). In another of his images of Hell on earth, he shows us Tityos, who in Greek myth could not control his desires—literally his love—and was punished by vultures feeding on his liver for eternity; this is an image, he says, of our inability to control our desires now, in this life (3.992–1002). In short, the hope for personal immortality, held out by the modern project, prevents us from being happy. As Lucretius says, "You may live through as many generations as you wish: nevertheless, that eternal death will still be waiting" (3.1090–91).

PART III

Consequences, Critiques, and Corrections

Chapter 10

Jean-Jacques Rousseau: Return to Nature vs. Conquest of Nature

Arthur Melzer

Earlier essays in this volume have done a fine job of indicating what it was in the classical thinkers that may have disinclined them from turning in the direction of a project for the conquest of nature. But these essays did not and could not discuss what the ancients explicitly said and did in order to discredit and combat that movement—for the simple reason that, in their time, there was no such movement to combat. We would have to wait another twenty centuries before a thinker arose who lived in a new world dominated by the conquest of nature, but who largely returns in his thinking to the ancients, and so who expressly and vigorously opposes that movement. That thinker is Jean-Jacques Rousseau.[1]

Emile, Rousseau's most comprehensive philosophical work, begins with this unique declaration: "Everything is good as it leaves the hands of the Author of things; everything degenerates in the hands of man. He forces one soil to nourish the products of another, one tree to bear the fruit of another. . . . He mutilates his dog, his horse, his slave. He turns everything upside down, he disfigures everything; he loves deformity, monsters. He wants nothing as nature made it, not even man" (*Emile* 37). To be sure, most earlier thinkers had maintained much of what is asserted here: things are good when they are in conformity with their nature or with God's design; but, tragically, they fall short of that conformity. Yet, heretofore, this "fall" was attributed either to sin, in Christian thinkers, or to ignorance, in most classical thinkers. What is unique in Rousseau's declaration is that it places the primary blame on a drive to control everything, to master and alter nature. He thus becomes the first to thematize and problematize "the conquest of nature," to the point of portraying it as the central issue. Though his critique of technology is quite extreme, it is the clearest and arguably the most powerful that we possess. Furthermore, reacting against this phenomenon, he also became the first to promulgate a new nature-religion as the basis for a new countermovement of "return to nature." Rousseau, in short, was the thinker who put "the problem of technology," as it later came to be known, on the agenda of modern philosophy and culture. The echoes of his thought can be heard clearly in the Romantics of the following century, as well as in contemporary environmentalism and in Heidegger.

Yet, it is the common misfortune of influential thinkers to be continually misunderstood. Their thought gets assimilated to all the related but different ideas proffered by the many thinkers they influenced. In order to uncover the precise grounds and character of Rousseau's opposition to the conquest of nature, then, we need to proceed with caution, step by step, especially because Rousseau's thought is also, on a number of other grounds, notoriously complex.

There are, in fact, at least four grounds of Rousseauian complexity, which it will be useful to state at the outset so as not to lose our bearings going forward. First, Rousseau is a powerfully dialectical thinker, which leads to the co-presence of opposites in his writings. He may eagerly join certain thinkers—the philosophers of the Enlightenment, for example—in their fundamental motives and first principles and yet end up as the founder of the counter-Enlightenment because, thinking through these shared principles in a more rigorous or radical way, he draws opposite conclusions. Thus, the Enlightenment and the counter-Enlightenment (including a hostility to technology) co-exist in his writings, since the latter, in his view, is simply the former more consistently worked through.

Second, Rousseau never for a moment loses sight of the tragic fact that what is good for the individual (or certain rare individuals) is not necessarily good for society, and vice versa. Thus, every question has potentially two different answers, one from each of these very different points of view.

Third, as an Enlightenment thinker, Rousseau is never content simply to elaborate some abstract, theoretical ideal but is driven to then turn to the issue of practical application. In executing this turn, moreover, he is an eager disciple of Montesquieu, holding that the self-same institution can be either good or bad depending on the character of the political whole in which it appears. Thus, in the *First Discourse* and the *Letter to d'Alembert*, Rousseau loudly proclaims the theater a highly pernicious institution for healthy republics like Geneva. Yet, in the latter writing and elsewhere, he also quietly concedes that it is actually salutary for the world of large, decadent monarchies like France, which explains the apparent contradiction that the author of these famous declamations also published plays and operas (*d'Alembert* 65; *Narcissus* 196).

Fourth, Rousseau, again like most Enlightenment thinkers, is an esoteric writer, as he openly acknowledges. He appears to be willing, among other things, to promulgate certain doctrines to which he does not subscribe in the hopes of moving the world in a healthier direction (*Second Letter to Bordes* 184–185; *Reveries* fourth walk). I believe that such is the case with Rousseau's new nature-religion mentioned above, and that the awareness of this fact is the necessary starting point for any adequate analysis of Rousseau's position on our topic.[2]

With Rousseau's constant rhetoric of a divine and teleologically ordered nature that "does everything for the best" (*Emile* 80), readers understandably assume that this is the true source of his opposition to the conquest of nature. But this religious doctrine, I suggest, is not the cause but the effect of that opposition. If one looks beyond the rhetoric to the actual arguments that Rousseau makes in rejecting the project of conquest, one finds that they are not theological, but entirely political and psychological. He simply argues that the project conflicts with the healthy and happy life, both for communities and individuals. But to give this opposition more real-world power, he links it to a religion of nature, akin to what one finds later in the Romantics, which gives the conquest of nature the feeling of something unholy—and the return to nature, a sense of grateful reunion with the divine source.[3]

The Truth of Modern Science—and the Good of It

To approach Rousseau's real views on the subject of modern science and the conquest of nature, let us begin our analysis with Rousseau's Enlightenment beginnings. In his basic premises, Rousseau is clearly a man of the Enlightenment. As one easily sees in the *Second Discourse*, notwithstanding certain religious protestations, his approach to the study of man is unflinchingly scientific and anthropological, his epistemology is empiricist, his psychology is egoistic and reductionist, and his political theory is built on rights-based egalitarian individualism and social contract theory. And more radically than anyone before, he argues that man, in his true nature, is not fashioned in the image of God, nor is he Aristotle's rational animal; he is a poor, solitary, amoral, arational, sub-human brute. Similarly, in his motivations, Rousseau shares the Enlightenment's hope to liberate humanity from religious and political oppression and, thus, also shares with it its greatest enemy: the Church (and, secondarily, the monarchy).

Turning, with all this in mind, to Rousseau's view of the new natural science (but not yet to the question of the conquest of nature), it should come as no great surprise to learn that he was strongly impressed and attracted by it, and indeed, through most of his life, he engaged in scientific researches, primarily chemistry, during the 1740s (see his *Les institutions chimiques*) and botany thereafter (see his extensive botanical writings, which mostly follow the system of his Swedish contemporary Linnaeus). His enthusiasm for the new science is displayed very clearly in the *First Discourse* (62–63): when he seeks to name the thinkers who have raised "monuments to the glory of the human intellect," and "whom nature destined to be her disciples," he points, not to Plato or Aristotle, but to the great modern natural philosophers Bacon, Descartes, and Newton. Bacon, in particular, he calls "the greatest, perhaps, of philosophers."[4]

Of course, Rousseau also greatly admires and relies upon certain modern *political* philosophers, above all Hobbes and Locke, but he does not name them here, no doubt because he ultimately considers their thought radically defective and in need of being replaced by a fundamentally new political philosophy. But Rousseau never speaks of any corresponding deficiency in modern *natural* philosophy. He registers no protest against it, as later philosophers and poets will. He would seem to be a firm admirer of Baconian science and Newtonian physics. But if Rousseau regards modern science as essentially true, that conclusion by no means settles for him the questions of whether science is good, in what way, and for whom. Those further questions form the subject of his first writing, the *Discourse on the Sciences and Arts*, where he declares his great break with the Enlightenment. The thesis of that complex work is stated most simply in one of Rousseau's later replies to his critics. He makes three points:

> [1] If celestial intellects cultivated the sciences, only good would result. [2] I say the same of great men, who are destined to guide others. . . . [3] But the vices of ordinary men poison the most sublime knowledge and make it pernicious for nations. (*Final Reply* 111; see 88)

Philosophical or scientific knowledge, pursued for its own sake (not for any practical benefits), is the highest good for the few genuine philosophers living the "celestial," contemplative life. But it is simply bad for ordinary men, not only because, misunderstanding it, they stumble from one sophistical idea to another, but also because it is inimical to true citizenship. The lofty contemplative life—detached, cosmopolitan, gentle, and self-concerned—undercuts the ancient Spartan ideal of the committed, patriotic, war-like, and self-sacrificing republican citizen. And fervent adherence to this ancient republican ideal is essential, in Rousseau's view, for any genuinely healthy political life. No citizens, no city. And yet, on the other hand (the third), philosophy need not remain wholly irrelevant to the people but can, in an indirect way, bring moral and political benefits to them through the intermediation of the "great men, who are destined to guide others." Such men are the legislators and, to a lesser extent, book-writing philosophers who, while scorning the Enlightenment effort to disseminate philosophy directly to the people, use their philosophic understanding to design and propagate among the people salutary institutions, laws, beliefs, and religions: "Greece owed its morals and its laws to philosophers and legislators. I acknowledge that. I have already said a hundred times that it is good for there to be philosophers provided that the people don't get mixed up in being philosophers" (ibid. 115). Rousseau's complex view is essentially the classical one: the highest good of science and philosophy is neither practical nor technological but contemplative and, so, the preserve of a few "celestial intellects." It is corrupting to the people. But certain philosophic legislators can nevertheless benefit the people by helping to make them, not contemplators and not technologically empowered consumers, but virtuous and patriotic citizens.

Thus, while Rousseau greatly admires the advances in our understanding of the physical world achieved by modern natural philosophy, he nevertheless rejects the account given by modern political philosophy of the role that this scientific knowledge should play in the political world—including its use for the technological conquest of nature. To understand more precisely what Rousseau is rejecting, and on what grounds, let us turn to a quick description of the Enlightenment as he saw it.

The Value of Technology in the Two Alternative Forms of Society

In an oft-quoted sentence from the *First Discourse* (51), Rousseau points to a difference between ancient thinkers and modern: "Ancient political thinkers incessantly talked about morals and virtue, those of our time talk only of business and money." To spell out the sweeping view implicit in this brief statement: Rousseau holds that there are basically two opposite ways in which one can attempt to unite naturally selfish human beings in social cooperation. One can, like the ancients, bend every effort to instill within people public-spirited "morals and virtue," which, by severely restraining the selfish, acquisitive desires and heightening fellow feeling, unify people through genuine respect and affection.

Conversely, one can attempt, like the moderns, to produce sociability from selfishness itself. The more one unleashes human selfishness and acquisitiveness, after all, the more people will feel the need of things, and the more they need things, the more they will require the help of others to supply their needs, and the more they depend on others, the more they must be willing to serve others so that these others will serve them in turn. In this way, selfishness, through a kind of inner contradiction, generates sociability.

The modern Enlightenment thinkers, beginning from the hard-headed premise that humans are not really the social animals Aristotle had claimed them to be (they are asocial and selfish, rather), switched their allegiance to the second alternative as a more natural, realistic, and elegant solution to the political problem. In light of this fundamental political dualism, the question before us needs to be restated as follows: in what ways are science and technology useful or dangerous for society—in each of its two alternative forms?

Well, in the new world of the Enlightenment, which no longer has an essential stake in strict republican morals, it was no longer dangerously corrupting to disseminate philosophical knowledge among the people. On the contrary, popular enlightenment now became positively useful because the new science, through its uniquely rigorous and verifiable explanations of natural phenomena, could help, as never before, to push back the kingdom of darkness and superstition. Furthermore, this science was (or could be) not merely theoretical and contemplative but also practical, in the new sense of technological: it promised increased power over nature for the "relief of man's estate," in the Baconian phrase. And this had a number of meanings. It meant, first, that people would experience in their own lifetimes the amelioration of various, seemingly permanent evils, thanks to new technologies. But, more sweepingly, it meant that the human race, in its age-old longing for salvation from the miseries of this life, might now find in the progress of its own scientific reason a new historical force, a secular messiah, on which to pin its heretofore millenarian hopes. Finally, it was not merely a question of material advance—both in the present and continuing programmatically in the future—but also a matter of human dignity and self-worth. Whereas, heretofore, human beings understood their self-worth in terms of their divine origin or their place in an ordered nature (both of which, however, made men's worth conditional on their humble obedience to these forces), henceforth they would find it in their god-like freedom from and mastery over nature. And this multi-leveled shift in life orientation would then feed back into and strengthen the new social order of selfish sociability by further liberating and legitimizing humanity's material concerns and desires.

Finally, if this new society was to work well in harmonizing people's selfish material interests, it was essential that it strive not for a traditional, static economy but rather a dynamic, growing one. Only when the whole pie is continually increasing does one person's gain not necessarily constitute another's loss. Machiavelli, for example, recommended constant imperialism as the only way of harmonizing the interests of the great and the poor (*Prince* chap. 16). But a superior driver of economic growth is the increase in labor productivity stemming from technological advance. In this way, the new technological science becomes important, not only for the provision of wealth and new conveniences, nor only as a secular displacement for our providential hopes, but also as the crucial source of dynamism and growth needed to sustain the whole new social order itself. It is the conquest of nature which, replacing the conquest of one's neighbors, makes it possible for selfish and acquisitive individuals to unite together harmoniously in a growing economy.

This, in brief, is the Enlightenment scheme that, according to Rousseau, all progressive writers of the time celebrated as the "masterpiece of the politics of our century." Rousseau himself, however, finds them horribly mistaken. "Of all the truths I have proposed for the consideration of the wise," he declares, "this is the most surprising and the most cruel" (*Narcissus* 193). When it is claimed that "[modern] society is so constituted that each man gains by serving the others," Rousseau objects, "This would be very well, if he did not gain still more by harming them. There is no profit, however legitimate, that is not surpassed by one that can be made illegitimately, and wrong done to one's neighbor is always more lucrative than services" (*Second Discourse* 194–195; see *Narcissus* 193). Rousseau does not deny that the modern, acquisitive individual will often be driven by his very selfishness to conclude cooperative agreements; and these will indeed be "mutually beneficial" in that they will leave all parties better off than they were before—a win-win situation. But all this does not suffice to make them truly in people's selfish interest. For each party to the agreement sees very clearly that he would be better off still if he received these same benefits without doing his share, and it is to the achievement of this optimal outcome that the strong will devote all their strength (and the weak, all their ruses). One can, of course, object that the government will outlaw cheating and exploitation, but how effective is that likely to be when the ill is so universal, and when the same selfish motives must be assumed to infect the government as well.

In Rousseau's radical view, then, the Enlightenment scheme of selfish sociability can succeed in producing the appearance of cooperation but never its reality. Prefiguring the Marxist Left, he sees such a society as a conflict system, a community of smiling enemies, drawn and held together by their need to use each other. It simply organizes us for mutual deceit and exploitation (*Second Discourse* 194–195).

Rejecting the Enlightenment on these grounds, Rousseau returns—with a vengeance—to the opposite alternative, the classical martial republic. And both of these moves, the rejection and the return, furnish Rousseau with good reasons for opposing the scientific conquest of nature. Where the Enlightenment celebrates technological advance for helping to make the new society of acquisitive selfishness possible, Rousseau opposes it for precisely the same reason. It is an essential cog in a destructive machine.

Conversely, in turning to the ancient martial republic, Rousseau fervently embraces, as he must, everything that suppresses selfishness, promotes material austerity, hardens and invigorates men, and attaches their purified, energized spirits to the city and to virtue. Viewed from this political standpoint, the project for the scientific conquest of nature is counterproductive, to say the least. Through the material ease that it fosters, the softening comforts and conveniences it introduces, and the larger materialistic hopefulness that it inspires, it powerfully subverts the austere and self-abnegating Spartan ethos.

In sum, Rousseau's view of the modern project for the conquest of nature is that it is politically harmful in the two most essential respects: it fatally corrupts and undermines the good form of society based on "morals and virtue," while being a major enabler of, and false advertisement for, the bad form based on "business and money."

Broadening the Question

This two-pronged political objection to technology is what first comes to sight in Rousseau's works. But, while it is an important part of his view, it is not the whole of it. It needs supplementing on at least two grounds.

First, this objection has been formulated from a very specific perspective—that of society, of what is needed in order to make political communities healthy or to make the individual good for others. But this is not necessarily the same as what makes individuals good for themselves. According to Rousseau, many things that are bad for Sparta are good for a private individual such as Emile, or Rousseau himself. Examples include: romantic love, isolated family life, softness and sentimentality, the theater, and philosophy. Thus, it is essential to evaluate technology all over again, now from the possibly very different standpoint of its consequences for individual happiness.

Second, we also need to follow Rousseau in his wonted shift from the standpoint of abstract theorizing to the question of what, in practice, is most beneficial under prevailing historical circumstances. In Rousseau's strongly pessimistic view, the theoretical ideal, the healthy Spartan republic, can no longer be restored in the historical conditions of the modern world, with its large states that can only be ruled monarchically, its tastes and morals that are irreversibly decadent, and its religion—universal, other-worldly, and gentle—that is antithetical to the republican spirit (*Social Contract* 126–130; *Emile* 40). Only a few distant approximations to the genuinely healthy community, such as his birthplace Geneva and a few other small republics, still manage to linger on in his time. Rousseau's practical purpose in proclaiming his passionate republican ideal is primarily to help them: to give these endangered citizens clearer and more rigorous foundations for their republican principles, and to aid them in reforming and strengthening their political institutions.

But for the world of large, decadent, and despotic monarchies, Rousseau had no serious hopes on the political level. He certainly proclaimed his theoretical republican ideal there, too, but his practical purpose was the opposite: not to encourage republican revolutions but, rather, resignation and detachment. Rousseau makes this perfectly explicit in *Emile.* When Emile—a Frenchman—is about to get married, his tutor suddenly uproots him and takes him on a long journey in quest of a legitimate and nonoppressive regime in which to settle with his new family. To theoretically inform this quest, he also steeps him in the teachings of the *Social Contract.* Rousseau states the intended effect of this theoretical study, telling Emile, "You will be cured of a chimera. You will console yourself for an inevitable unhappiness, and you will submit yourself to the law of necessity" (*Emile* 457–458). In the modern, monarchical world—a new age of iron—the public realm is hopelessly illegitimate and oppressive. One needs to understand and adjust to that fact.

But one consequence of this dire practical situation is that one is released by it to save oneself, to pursue a private, individual happiness, especially—as Emile will do—by fleeing the cities (the epicenters of bourgeois, selfish sociability) for rural isolation, and by retreating into the bosom of the family: the new, love-based, child-centered, sentimental family (of which Rousseau was the first and greatest ideologist).

In short, it turns out that, at this particular moment in history, the two shifts in perspective called for above essentially point in the same direction. If we turn from the abstract, theoretical ideal to what is required in practice under the historical conditions of the modern, monarchical world, we will, at the same time, be making the needed turn from the standpoint of society's good to that of the private individual. So, the question before us now becomes: what benefits or harms does technology—regardless of its consequences for society in its two alternative forms—hold for the individual who seeks a private happiness outside or apart from society?

Technology and Individual Happiness

To state Rousseau's answer in a word, the technological project is pernicious for the individual for essentially the same reason as it is for the citizen: it conflicts with and undermines *moderation*—that word, so profound and meaningful for the ancients, so shallow and boring for us. Of course, it is obvious, even to us, why the Spartan citizen needs moderation: he is called to subordinate his own good to that of others. But what is the need for such austerity and self-abnegation where the issue is solely one's own well-being? Moderation obviously makes you good for others, but how does it make you good for yourself? That is the central question here.

Rousseau would approach it as follows. "In what," he asks, "consists human wisdom or the road of true happiness?" (*Emile* 80). He proposes a simple schema. We have our needs and desires on the one hand, our faculties for satisfying them on the other. "Our unhappiness consists, therefore, in the disproportion between our desires and our faculties. A being endowed with senses whose faculties equaled his desires would be an absolutely happy being." While it is not possible for human beings to equalize these two things completely, our goal should be to minimize the gap between them as much as possible.

But if that is the essential task of life, then two fundamentally opposite life strategies present themselves: we can either restrict our desires or extend our powers—self-mastery or world-mastery. And between these two choices, the life of austere moderation and the life of conquest, Rousseau sees no real choice: only moderation can succeed. Conquest, which might seem at first the superior path, turns out to be self-defeating, Rousseau will argue.

Therefore, he embraces moderation—but not because of any pious restraint before holy nature. Also, not because he is a puritanical Christian who condemns desire as sinful. On the contrary, he never tires of proclaiming that, prior to our corruption in society, human beings are naturally good, and all our basic desires are innocent and healthy (*Emile* 212). For the same reason, Rousseau is no Buddhist ascetic equating desire as such with suffering and calling for its complete extirpation. He goes out of his way to caution that if our desires were diminished "beneath our power, a part of our faculties would remain idle, and we would not enjoy our whole being" (*Emile* 80). Rousseau embraces the moderate life, the simplification of existence, purely because he finds it to be the true art of living, the only path that offers one profound peace, pleasure, and satisfaction. Not unlike the ancient Epicureans, Rousseau believes in a kind of hedonistic or sensuous asceticism.

By contrast, the alternative strategy—the quest for power, whether economic, political, or technological—is self-defeating in at least three different ways. People always imagine, to take the most common example, that if only they had ten times as much money as they do now, they would have all they desire. The flaw in this common fantasy, however, is that you always think of yourself and your desires as remaining unchanged—you hold that part of the equation constant—so your newly increased wealth and power strike you as more than enough. But in fact, great wealth changes you, and the little things you now desire will not continue to satisfy you then. Your relative poverty, which you rebelled against, in fact had a salutary, moderating effect on you, which has now been abruptly removed. And desire, if liberated from all control, whether from external constraints or internal discipline, has an inherent tendency to grow limitlessly. It grows because the things you long for eventually disappoint you when you acquire them, pushing you on to new desires. The disappointment grows partly from the numbing effect of habitual possession (*Second Discourse* 147). But also from this: your imagination richly adorns and idealizes the things you long for, but it abandons them when you have them securely in your hands. Then, alas, they simply are what they are. Only absence makes the heart grow fonder. Thus, most human desires and passions, while not sinful or wicked in Rousseau's view, are inherently deceitful. They always promise more than they can deliver. And so long as this recurring disappointment does not finally teach us the untrustworthiness of desire and the wisdom of moderation, it will push us to a still more determined pursuit of new goods—the *real ones* this time—and so on without end (*Emile* 80–81, 242–243).

The effort to achieve an equality of desire and power through the pursuit of the latter is self-defeating, then, because as power increases, so does desire, and at a far faster rate. For power must be rooted in the real world, which has its laws and limits, whereas desire stems from the imagination, which knows no limit. As we have heard since ancient times, desire is infinite. In more contemporary terms, as economic and technological power increase, they inevitably produce not only new and more goods, but new and more needs.

There is also a second way in which the quest for power is self-defeating. It not only stimulates desire but undermines power itself. In seeking power, for instance, we acquire things; in acquiring things, we extend ourselves; in extending ourselves, we increase our exposure and insecurity, hence, our need for power. It is this inner contradiction that turns the limited desire for power into the ceaseless quest that Hobbes famously speaks of. The cause of this, as Hobbes explains, is "not always that a man hopes for a more intensive delight, than he has already attained to; or that he cannot be content with a moderate power: but because he cannot assure the power and means to live well, which he hath present, without the acquisition of more" (*Leviathan* XI, 80).[5]

Rousseau agrees with this Hobbesian point, only spelling out more fully its consequences. He particularly emphasizes that even those at the pinnacle of power, absolute tyrants, are afflicted by this problem: they are slaves of their power, spending all their days trying to protect it. As Rousseau states, "Whoever is master cannot be free, and to reign is to obey. Your Magistrates know that better than anyone, they who like Othon omit no servility in order to command" (*Montagne* 841–842; see *Second Discourse* 173). Many a workaholic CEO of our day must have a good sense of how this feels. The tyrant appears to be free and powerful, since he constantly makes others do what they do not want to do; but to acquire, protect, and exercise this vast "power," he must constantly do what *he* does not want to do. This is the meaning of the famous and puzzling declaration that Rousseau places at the beginning of the *Social Contract*: "One who believes himself the master of others is nonetheless a greater slave than they." This is the classic statement of the self-defeating character of power.

Rousseau also points to a somewhat different way in which our increasing power, especially technological power, can weaken us. Speaking admiringly of the strength, vigor, and toughness of savage man, he states, "If he had an axe, would his wrist break such strong branches? If he had a sling, would he throw a stone so hard? If he had a ladder, would he climb a tree so nimbly? If he had a horse, would he run so fast?" (*Second Discourse* 106) Human progress from savagery to civilization, as Rousseau recounts it in the *Second Discourse*, is largely the story of the strengthening of our technologies and the weakening of ourselves, of our inner strength and abilities. For technology essentially consists in the outsourcing of our faculties. No doubt we gain access to various enormous powers in this way; nevertheless, as our tools are perfected, we ourselves atrophy. Indeed, sometime in the relatively near future, there will be smart machines that can do a far better job of literally everything that human beings can do with their bodies and minds. What, then, will be left for us to do? And what, then, will we become?

If all that matters in life is what you have, and not what or how you are, then the atrophying of our faculties and energies will be no great loss. But if, as Rousseau suggests in a passage already quoted above, the good life requires the employment of all our faculties and the exercise of our inner strength, so that we "enjoy our whole being," then the vast growth of our technological power will, in the decisive respect, leave us weaker (see also *Emile* 42).

Finally, there is a third way—the most fundamental—in which the strategy of conquest proves utterly futile. As Rousseau acknowledges, even under the best of circumstances, we can never completely close the gap between desire and power. What he has in mind, of course, is that we humans, being vulnerable and mortal, necessarily desire to escape that condition, while necessarily lacking the power to do so. Thus, our true task as human beings is to narrow the gap between desire and power as much as possible and, above all, to find some way to live with the inescapable part of the gap that remains.

This ultimate task is clarifying. In its light, it becomes perfectly clear that the pursuit of power, perhaps useful for other things, is of absolutely no help in the decisive respect. There is no amount of power that will take you even one tiny step in the direction of immortality. (Our increases in longevity, while certainly a good thing, always fall equally, because infinitely, short of immortality.) From all the writers of antiquity, we learn that the only (secular) solution is to be found in the opposite strategy, in moderation in the deepest sense of that word: knowledge and acceptance of man's unconquerable limits, resignation to necessity, stopping short before what cannot be changed. That is what it means to learn how to die, which, Rousseau insists, is the true key to learning how to live. If you don't know how to die, then you will spend your whole life obsessively, if largely unwittingly, in the futile effort to flee or resist it, with the result that you will die but without ever having really lived. You will continually ignore, postpone, and distort your existence, busying and burdening yourself with the impossible task of preserving it. As Rousseau puts it in *Emile*: "I am not able to teach living to one who thinks of nothing but how to keep himself from dying" (53). Only in accepting the necessity of death do you stop wasting the finite life that you have been given—and start to live. For you can only truly love and enjoy what you know how to lose. But this takes enormous strength.

Rousseau's whole education of Emile is designed with this thought in mind. Thus, it is above all an education in strength and hardness, achieved through a training in adversity. As he explains:

> What does not admit of exceptions . . . is man's subjection to pain, to the ills of his species, to the accidents, to the dangers of life, finally to death. The more he is familiarized with all of these ideas the more he will be cured of the importunate sensitivity which only adds to the ill itself the impatience to undergo it. The more [Emile] gets used to the sufferings

> which can strike him, the more, as Montaigne would say, the sting of strangeness is taken from them, and also the more his soul is made invulnerable and hard. (*Emile* 131)

From earliest youth, a person must be confronted with hardship, difficulty, and suffering, not in order to preoccupy him with evil and the effort to avoid it, but precisely to cure him of that futile preoccupation and thus free him for a positive orientation in life, a life spent pursuing what is good and not obsessively fleeing what is bad. This training in adversity will work in at least two ways.

First, on the intellectual level, Emile will learn what life is really about. By repeatedly slamming into the hard walls of the world, he will wake up from the fantasy of immortality that we all gravitate to. It will be a kind of reality therapy, constraining him to live in the real world. He will acquire a wise disillusionment and a salutary fatalism. He will learn, in his bones, the harsh, inescapable truth of the human condition: that the story of every human life has the same ending—the hero dies. He will steel himself to this fact, holding himself to a posture of acceptance and endurance, not resistance.

But to hold unflinchingly to such a harsh vision of life, one that contradicts all our deepest longings, requires tremendous inner strength. Thus, beyond the purely intellectual component, the training in adversity also aims to toughen and harden him both physically and psychologically. It will give him the sense that whatever evils the future will bring, they are not beyond his ability to endure; he is ready for them—a strength and confidence that will free him to unclench his fist, to live in and enjoy the sentiment of his existence, the positive good of being alive this moment, even while awaiting the inevitable end.

If this training in adversity is, for Rousseau, the true core of an education for happiness, then it is obvious that the pursuit of power in general, and the technological conquest of nature in particular, are not only the wrong life-strategy—being powerless to address the central issue of life—but are destructive of the right one, the path of moderation and resignation to necessity. They directly undermine the two purposes of the education in adversity we have just seen.

The growing comfort and softness of life provided by technology obviously works directly contrary to the second purpose, the toughening and hardening education that Rousseau sees as so crucial. Indeed, our world has already become so soft that the whole concept of a toughening education has lost its rationale. Once upon a time, we all readily acknowledge, life was hard and men had to be strong. But thanks to technology, we have conquered and tamed life, so what possible need is there for this old-fashioned insistence on adversity and strength? It strikes us as palpably unhealthy, either fascistic or masochistic.

Rousseau would respond that the importance of human strength has always had less to do with our external challenges and more to do with the internal one. The defining struggle of life concerns not our relation to the world but to our own mortality. And despite centuries of dazzling political and technological progress, nothing has changed on that decisive front. Nothing. To live our lives, today, in a free and positive way, in spite of death, requires every bit as much strength as it did in the past (if not more)—and, indeed, a great deal more than most of us can muster.

But, as we have seen, in addition to the training in strength, Rousseau's education in adversity also has a second, intellectual purpose: to teach Emile to see life as it really is, to "return to nature" in the deepest sense of living in the true world, to understand the harsh limits of the human condition and accept them without futile rebellion. This is not to deny, of course, that life is also full of a multitude of evils that can and should be resisted and overcome. There are many useful arts. But these successful conquests, though real, must always be kept in proper perspective. Most things in life can be changed; some few cannot. But mortality, the master fact of human life, falls squarely in the latter category. Thus, in any proper vision of human life, the value of conquest is strictly subordinate to that of resignation. If there is any *fundamental* help for us, it is through resignation.

But the technological conquest of nature—seen not merely as a series of useful inventions, but as a project and a worldview—subverts this whole intellectual education, indeed stands it on its head. It teaches the fundamental priority of conquest to resignation. It preaches not a salutary fatalism but what can be called solutionism, the belief that there are no problems which the human race cannot eventually solve. Thus, it tends not merely to reject the belief in necessary limits but to debunk it as cowardice and as self-fulfilling defeatism. Conversely, it looks upon the posture of rebellion, the resolute refusal to recognize natural limits, as something to be celebrated; indeed, as the highest virtue and the key to the future.

It must be confessed, of course, that it is hard not to be inclined to such a view when we continually see things once regarded as impossible fall before the juggernaut of our science. Our constant experience is of the continuing and glorious expansion of the realm of the possible, the realm of freedom. But still, if it is our aim to see the world as it is, we must also confess that, according to the laws of physics as currently understood, it does not seem that our ever-evolving universe can remain supportive of life eternally. So even under the most imaginative schemes for longevity, we remain mortal and thus continue to labor under the paramount necessity of adjusting ourselves to that fact.

In sum, moderation, in the deepest sense of knowing and accepting our limits and learning how to die, is the central task for us, but it is by nature and under all historical circumstances extraordinarily difficult, since our every inclination pushes us toward preservation. But in our age, it has become even more difficult due to the modern project for the conquest of nature that coddles and softens us through its material effects at the same time as it misleads and corrupts us through its guiding assumption that there are no natural limits for us and that resignation to necessity is an attitude toward life that is fundamentally outmoded, cowardly, and counter-productive.

One Final Twist: The Road to Kant

We have seen that Rousseau opposes the conquest of nature, not from the standpoint of any kind of nature-piety, but because it conflicts with the virtue of moderation that is necessary for the healthy and happy life, both of communities and of private individuals. But this statement requires some qualification. It is true if "the conquest of nature" is taken to mean what it ordinarily means and what I have been using it to mean: the control of external nature for the sake of man's material needs and acquisitive desires.

But Rousseau specializes in taking Enlightenment concepts and institutions and turning them to his own, often counter-Enlightenment purposes. That is true here, too. Thus, when Rousseau proclaims, in opposition to the Enlightenment, that a healthy society must strictly moderate men's acquisitive passions, replacing them with an ardent patriotism, he himself makes use of the Enlightenment concept of conquering nature, albeit in a different sense. For, since Rousseau agrees with the Enlightenment that human beings are naturally asocial, one can produce the needed patriotic citizens only through a conquest of human nature. Thus, describing the great Legislator in the *Social Contract* (68), Rousseau states: "One who dares to undertake the founding of a people should feel that he is capable of changing human nature, so to speak." Again, in *Emile* (40): "Good social institutions are those that best know how to denature man."

And it is not only the Legislator, in Rousseau's account, who will be engaging in the conquest of human nature but also (if somewhat inconsistently) the virtuous citizens themselves. In the *Social Contract* (56), he gives a famous description of the citizen's "moral freedom, which alone makes man truly the master of himself. For the impulse of appetite alone is slavery, and obedience to the law one has prescribed for oneself is freedom." In the *Second Discourse* (114) as well, Rousseau suggests that man's true uniqueness and dignity is that, through morality, he rises free and superior to the mechanistic laws of nature.

Rousseau is always very tentative in his discussions of this new notion of moral freedom, and it is quite possible that here, as with his religion of nature, he is more persuaded of the utility than the truth of this doctrine. But whatever the case, it remains remarkable—and very important for his influence on later, especially German thought—that Rousseau does not leave it at simply rejecting the modern project for the conquest of nature as something destructive of moderation, as described above. In certain contexts, he also goes on to adopt and dialectically transform that seductive modern conceit of man's god-like freedom and mastery into something that serves his own, quite opposite goals. He turns the conquest away from external nature and back onto man himself—human self-conquest. Indeed, he directs the conquest against that part of man that greedily seeks to conquer external nature. In so doing, he reverses everything: he transforms man's conquest into a powerful new force for moderation itself. In short, Rousseau embraces (or at least makes use of) the modern, humanistic view that man's true dignity consists not in obeying, but in rising above and mastering the realm of nature, but he reinterprets this as the mastery of one's own natural appetites through moral freedom. He turns the technological conquest of external nature into Kantian autonomy.

Chapter 11

Kant on Organism and History: Ambiguous Endings

Richard Velkley

Philosophy as Relief of the Human Condition

Immanuel Kant's philosophy has a qualified endorsement of a version of the idea of mastery of nature even as it offers some of the most trenchant criticisms of the idea.[1] I shall borrow a few words of Leo Strauss for a pithy summation of the idea, in a formulation that clearly has some bearing, albeit equivocally, on Kant's thought. Writing of Machiavelli, Strauss states: "He achieves the decisive turn toward the notion of philosophy according to which its purpose is to relieve man's estate or to increase man's power or to guide man toward the rational society, the bond and the end of which is enlightened self-interest or the comfortable self-preservation of its members. The cave becomes the 'substance.' By supplying all men with the goods which they desire, by being the obvious benefactress of all men, philosophy (or science) ceases to be suspect or alien."[2] Strauss proposes that the "realism" of Machiavelli, or his critique of the ideal republics and principalities of pre-modern philosophy, is adopted by the succeeding tradition of modern philosophy in its project of relieving man's estate, with the intent to overcome the philosopher's ancient plight as outsider and exile. Strauss ascribes to the modern philosopher the primacy of a practical end that is beneficial to all humans, although perhaps above all to the philosopher. The practical end surpasses, if it does not simply replace, contemplative activity as the core of philosophy. In another place, he writes that, "from the very beginning," modern man's "attempting to be absolutely sovereign, to become master and owner of nature, to conquer chance" issues in "the oblivion of eternity, or, in other words, estrangement from man's deepest desire and therewith from the primary issues."[3] By this account, the task of mastering nature requires the narrowing of the horizon of philosophical thought to the temporal-historical realm of human projects. Mastery of nature entails, already in Hobbes, historicist modes of thinking.[4]

No philosopher declares more forcefully than Kant that the end of philosophy is universal human improvement, wherein philosophy offers guidance of the human species toward the rational society. Although Kant speaks of the increase of human power through the development of skills in mastering nature, he says much more about goals of human self-mastery, with a moral inflection that seems distant from Machiavelli. All the same, the epigraph he takes from Bacon's *Great Instauration* for the second edition of the *Critique of Pure Reason* implies advocacy of the Baconian project of the relief of man's estate.[5] The Baconian idea of progress acquires a moral tone in Kant's hopeful speculations on human history construed as a story of gradual enlightenment and unending advance toward moral perfection. Provisionally, one can say that Kant moralizes the concern with the relief of man's estate, or with progress, mastery, utility, and power, as he remains within the modern framework of dedicating philosophy to human practical goals. Through this moral transformation, the bond of society ceases to be enlightened self-interest or comfortable self-preservation. The bond is now the moral law, understood as the categorical imperative, as well as notions of lawful order (external freedom as secured in particular states) that in complex ways are subordinated to the universal moral law.

Kant made the turn to his moral conception of philosophy's end through the inspiration of reading Rousseau, who, Kant eloquently announces, changed his view that the thirst for knowledge "alone could constitute the honor of mankind," thus setting him "upright," so that now he learns "to honor human beings" and regards the consideration of how "to establish the rights of mankind" as bestowing worth on his activity, which otherwise is "far more useless" than the common worker's.[6] Yet Rousseau did not, like Kant, place happiness below obedience to the moral law, and Rousseau's central concern with distinguishing between the ways of life of the citizen, the private individual formed by natural education, and the solitary thinker, seems alien to Kant. And certainly far from Rousseau is Kant's conception of human reason as progressing, with assistance from nature, toward a satisfactory completion of the powers and strivings inherent in its structure. This conception is the "guiding thread" Kant brings to the reflection on human affairs, the idea of a "hidden plan of nature" which supports "a comforting view of the future, one in which we represent from afar how the human species finally works its way up to the state where all the seeds nature has planted in it can be developed fully, and in which the species' vocation here on earth can be fulfilled. Such a *justification* of nature—or, better, of *providence*—is no unimportant motive for adopting a particular perspective in observing the world."[7]

The Telos of Reason's Self-Completion

The analysis of Kant's view of the progress of reason (his version of philosophy's coming to the relief of man's estate) must consider not only the moral end that Kant ascribes to such progress (the species' achievement of autonomy accompanied by a fitting natural satisfaction or happiness) but also comment on the peculiar teleological account of reason that supports the hope for attaining the end. Teleological thinking is central to the abstract metaphysical ruminations of the *Critique*, for in this work Kant defines philosophy as "the science of the relation of all knowledge to the essential ends of reason" and asserts that the highest end is moral.[8] He pronounces that the criticism of reason's theoretical powers is undertaken for "the interests of humanity": "I have therefore found it necessary to deny *knowledge* in order to make room for *faith*. The dogmatism of metaphysics, that is, the preconception that it is possible to make headway in metaphysics without a previous criticism of pure reason, is the source of all that unbelief, always very dogmatic, which wars against morality."[9] In bringing about the "end of infinite errors" (Bacon) as no earlier philosophy had done, criticism of reason serves human practical-moral ends, but it also answers a puzzle about the nature of reason, since the failure of the philosophic tradition to attain certainty in metaphysics, reason's deepest inquiries, poses several important questions.

> Why, in this field, has the sure road to science not hitherto been found? Is it, perhaps, impossible of discovery? Why, in that case, should nature have visited our reason with the restless endeavor whereby it is ever searching for such a path, as if this were one of its most important concerns? Nay, more, how little cause we have to place trust in our reason?[10]

> Reason is impelled by a tendency of its nature to go out beyond the field of its empirical employment, and to venture in a pure employment, by means of ideas alone, to the utmost limits of all knowledge, and not to be satisfied save through the completion of its course in a self-subsistent whole. Is this endeavor the outcome of merely the speculative interests of reason? Must we not rather regard it as has having its source exclusively in the practical interests of reason?[11]

Thus, the critique's attempt to show that reason's striving for metaphysical or supersensible knowledge is moved by practical (moral) interests and is satisfied only by them ("the primacy of the practical"), would disclose a teleological character in reason, or an "organic" structure, in which no faculty or interest is in vain. Indeed, reason, at least in a fundamental respect, is promised full satisfaction:

> Pure reason, so far as the principles of its knowledge are concerned, is a quite separate self-subsistent unity, in which, as in an organized body (*organisierten Körper*), every member exists for every other, and all for the sake of each, so that no principle can safely be taken in *any one* relation, unless it has been investigated in the *entirety* of its relations to the whole employment of reason.[12]

> This science [the critique of reason] cannot be of any very formidable prolixity, since it has to deal not with the objects of reason, the variety of which is inexhaustible, but only with itself and the problems that arise entirely from within itself, and which are imposed upon it by its own nature, not by the nature of things which are distinct from it.[13]

> Metaphysics . . . is the only one of the sciences which dare promise that through a small but concentrated effort it will attain, and this in a short time, such completion as will leave no task to our successors save that of adapting it in a *didactic* manner, For it is nothing but the *inventory* of all our possessions through *pure* reason, systematically arranged.[14]

The morality of reason's unconditioned lawgiving emerges as crucial to this completion and is not meaningful apart from it. There is a rationale of a strange and lofty kind for Kant's focus on reason as autonomous that one cannot draw solely from the tradition of "ethics."[15] The relief of man's estate, transformed into a project that is centrally about metaphysical satisfaction, ascribes to morality a new metaphysical significance, and the entire realm of politics, religion, commerce, and the arts and the sciences is interpreted as furthering morality, understood as the source of this significance.

This conception of reason is complex and endlessly perplexing. I wish to highlight some major points.

(1) Kant's view of reason as a self-subsistent whole, wherein metaphysical interests are satisfied by reason's own activity, employs notions of final causality and living natural structure that seem at first pre-modern, whereas they actually move the modern idea of the self-grounding of reason (about which more soon) in a new, radical direction.

(2) To characterize this self-satisfaction as a kind of natural teleology, or as organic, brings about an explicit conflict with thinking about nature and the human in mechanistic terms, which terms (Newtonian physics) Kant validates only "within limits." The conflict of teleology and mechanism has only a problematic resolution by Kant, and it places an obstacle in the way of mastery of nature, insofar as this requires the belief that the principles of the mechanistic science of nature suffice to explain the totality of things.

(3) In spite (or because) of Kant's meliorism and the moralizing tenor of his philosophy, he does not, in the end, hold that the human should try to be "absolutely sovereign" or that reason is able to conquer chance. In his later writing on religion, Kant presents a doctrine of radical evil that explicitly rules out the possibility of human attainment of moral perfection, and, indeed, he characterizes the belief in such attainment as fanatical.[16] Famously, Kant declares that "from such crooked wood (*aus so krummem Holze*) as man is made nothing straight can be fashioned."[17] What is more, he states that the enduring tension between the demands of nature and those of freedom guarantees the continuing existence of philosophy.[18]

(4) Kant claims, as already noted, to settle with finality the status of reason, to make sense of its metaphysical disposition, and to bring an end to infinite errors. This entails a certain conception of reason as historical, capable of achieving an unprecedented completion.[19] For the sake of the "interests of humanity," Kant proposes a final reckoning with the questions naturally inherent in reason (*metaphysica naturalis*) concerning the whole and its grounds. But even as the critique purports to resolve the fundamental problems, it acknowledges the permanence of the questions.[20] Does this

not mean that the questions are not answerable to human reason's full satisfaction?

Mathematical Philosophizing: Its Greatness and Misery

Kant provides substantial indications of the critical method's provenance in earlier modern philosophy in the preface to the second edition of the first *Critique* (1787). The role of mathematics in the "intellectual revolution" of the "past century and a half since Bacon" is underlined. Kant finds a "new light" was already seen in Greek antiquity by the first person who demonstrated the properties of the isosceles triangle: the true method was not to "read off properties" but to bring out what was implied in concepts formed a priori by construction.[21] But it took centuries to see that this method could be applied to natural science, when natural philosophers (Galileo, Torricelli, Stahl) learned that "reason has insight only into that which it produces after a plan of its own, and that it must not allow itself to be kept, as it were, in nature's leading-strings."[22] In language echoing Bacon, nature is constrained "to give answers to questions of reason's own determining"; the mind relates to nature not as the attentive pupil but as "the appointed judge." At the same time, reason must accept whatever has to be learnt from nature, and not arbitrarily ascribe properties to it. But the guide of reason in its quest for knowledge is found in itself: the guiding idea is a unified, thought-out plan rather than "accidental observations" and "random groping." Knowledge crucially depends on unifying principles that cannot be drawn from the phenomena; Kant, in the argument of the *Critique*, rests these principles on the a priori forms of intuitions and categories employed in synthetic a priori judgments. The emphasis on a priori sources of unification evokes Descartes's mathematizing of the Baconian conception of laws of nature, wherein something immediately certain to the mind, but not ultimate in things, is elicited from experience and then employed as a foundational principle. Thereby, a system of homogeneous abstractions (laws) replaces the heterogeneity of beings.

In the Cartesian move, the Baconian empirical certainty (e.g., "heat"), widely distributed through nature, is replaced by the geometrical concept of body ("extension"), intuited by the intellect, with absolute universal application to bodies.[23] The condition for Descartes's epoch-making step, grounding the mathematical treatment of nature, is the reflection on method in the early *Rules for the Direction of the Mind* (*Regulae*) and the accompanying revolution of the analytic coordinate-based geometry. Seeking a comprehensive unification of the sciences that will allow for both the pleasure of contemplation and practical benefit, Descartes uncovered the possibility of unification in the natural ability of the mind to grasp truths ("intuitions") that are present to it apart from the senses. The first application of this approach was the use of the indeterminate magnitude "proportion" that can fuse geometry and algebra into a single new science.[24] In his further reflection, Descartes needed something beyond the capacity to grasp intuitive certainties; indeed, something to substitute for the lack of a natural guide in the mind's directing itself toward knowledge: a new organon or art to replace the traditional logic that relies on natural assumptions for premises.[25] Thus, the unifying certainties (proportion, extension) must be applied within a larger unifying framework that conducts the mind in a progressive way toward comprehensive understanding. Such enlargement of method results from noting that "perfection" in works of architecture, city planning, and legislation issues from a unitary source; that is, a single, master intellect.[26] This conception goes far toward the replacement of natural teleology by an artificial *telos*, but, in fact, the philosopher cannot do without guidance by natural "teachings" of pleasure, pain, and desire, which resist clarification by the new method, and which show that the new science cannot wholly leave behind the realm of "pre-scientific" experience.[27]

It is often said that Kant advances on "rationalist" thought such as Descartes's by seeing the necessity to give primacy to active synthetic judging rather than the intuition of innate ideas, but this misses the point that Descartes's conception of knowledge is from the start a "dynamic" and expansive vision: of reason as progressively unifying the whole of what is available to human apprehension. It is not merely the search for absolute certainties (such as "I think") that can withstand the radical doubt induced by an imaginary divine or demonic deceiver.[28] Nor is the point of things adequately put by saying that Descartes seeks the foundation of natural science, for this assumes a notion of "science" as a simply theoretical enterprise.[29] Descartes transforms the meaning of knowing (with crucial indebtedness to Bacon) into a process at once practical and theoretical, at the same time mastering the given for human benefit and disclosing an intelligible order.[30] Kant's account of reason cannot be directly derived from this Cartesian source. He starts with the acknowledgement of the achievement of the new mathematical sciences in the Newtonian version that contains principles undiscovered by Descartes, and, furthermore, Kant's account of the foundations of natural science critiques some features of Newton on both metaphysical and mechanical levels.[31] Kant's central endeavor is to establish metaphysics itself as a science, since the "dialectic" generated by reason remains unresolved, which is partly due to modern mathematical science's not addressing reason's inevitable interests in questions ("ideas") about the whole and its grounds. Indeed, in this regard, the new mathematical philosophy is an abject failure.

Since mathematics can realize its concepts in intuition constructively, "it becomes, so to speak, master of nature (*Meister über die Natur*)," and its success gives rise to the expectation that it can have the same success in fields other than quantity.[32] But the pure concepts of the understanding (such as cause and substance) cannot be deduced by mathematical methods; their reality cannot be intuited a priori. These concepts, as conditions for the objectivity of all empirical knowledge, require a special deduction. Outside its sphere of application to appearances, mathematics is groundless, unable to stand or swim, and reason falls into chaos in the transcendental use of mathematical concepts.[33] Thus, the "uncritical" modern account of knowledge as mathematical mastery of nature fosters, rather than prevents, a natural state of injustice and violence in the controversies of dogmatic reason. Citing Hobbes, Kant demands that one leave such a state "in order to submit oneself to lawful coercion which alone limits our freedom in such a way that it can be consistent with the freedom of everyone else and thereby with the common good."[34] The solution to reason's internal discord is an a priori legislation that substitutes for natural teleology, again recalling Cartesian method, but it is a project focused on correcting reason's theoretical "injustices."

The internal coherence that results, however, seems to be not merely legislated, and not just an artifact, but an organism of a self-causing, self-sufficient sort. The language of reason's self-articulation links legislation and art to organism (perhaps via organon).[35] The centrality of teleology in Kant's investigation of reason leads him to the critical examination of purposiveness, with the internal division of the inquiry into two parts, aesthetic judging (including fine art) and living nature.

The Mystery of Organism

Kant completes his critique of the powers of reason with the *Critique of the Power of Judgment*, the third of the *Critiques*. Among the three higher cognitive powers—understanding, reason, and judgment (sensibility being a lower power)—judgment is peculiar in having no legislative domain (*Gebiet*) of its own within the broad territory (*territorium*) of possible objects of cognition, unlike understanding and reason, which legislate in a priori fashion over nature and freedom.[36] At the time of the first *Critique*, Kant did not conceive judgment as a faculty with its own a priori principle but as the power of subsuming given particulars under rules provided by understanding, the faculty of concepts. There can be no rules for subsuming under rules (under pain of infinite regress), and, hence, there is no special principle for judgment as limited to subsuming.[37] In connection with taste and aesthetic judging, Kant conceived of an a priori form of judging and, hence, a form suitable for critical inquiry.[38] The conception was broadened to include the general principle of purposiveness (*Zweckmässigkeit*), the judgment that something has an appearance of being purposive for human cognitive powers (as in aesthetic judging) or in its inner possibility (as in organism and nature as a whole as teleological). The general principle for these cases is judging as the search for universals where only particulars are given, and pre-existing, determinate universals are lacking. On behalf of finding such universals, judging "reflects" (i.e., assumes the presence of) a design or regard for the human faculties in given appearances that furthers the search.

This principle can be restated as the assumption of favorable conditions (causality by design) for the search for laws, which are contingent with respect to pure transcendental laws (the universal homogeneous laws of nature grounded in the legislation of understanding). The a priori principle of judging is not "constitutive," it cannot be ascribed objectively to appearances, and is only regulative, a requirement of human faculties to favor their own activities. In the case of aesthetic judgment (taste), a given form in sensory intuition initiates a harmonious play between imagination and understanding that is propitious for cognition, but which is enjoyed contemplatively in "mere reflection" without regard for knowledge, and judged beautiful. Yet Kant accords great importance to this principle, since reflective activity in both the aesthetic and teleological realms makes possible a transition (*Übergang*) from nature to freedom by proposing a supersensible substrate that underlies both.[39] Such unification of domains serves the interest of reason in discovering grounds for regarding the realization of freedom in the sensible world as possible. To promote this interest is what Kant adduces as the ultimate systematic rationale for the critique of the power of judgment and its principle of purposiveness, for such critique is the necessary propaedeutic to articulating the system that unites theoretical and practical employments of reason into a whole.[40]

This treatment of judgment exposes a strikingly ambiguous account of the human situation in nature. On the one hand, the perspective is wholly anthropocentric: the human faculties make assumptions about natural appearances favorable to human ends, and, at the highest level, they assume a design in the common ground of nature and freedom that promotes the practical interest of the human as an autonomous (self-determining) being. But Kant's account of the inquiries into the features of nature (organism) that appear to our faculties to offer the evidence of such design emphasizes the inherent, permanent inscrutability of those features. This poses not only a limit on what can be understood in these inquiries, but it banishes the hope of complete mastery of nature by means of the mathematical-mechanistic account of natural laws. It would be wrong to describe Kant's position as only a reductive one: teleology is merely a subjective, self-serving human interpretation of phenomena. Instead, he argues that purposiveness in nature is a permanent problem, in that living beings *are* structured in a way that baffles human comprehension and prevents the establishment of a biological science—inclusive of applications of the science that would enhance human power. In fact, Kant's stress on the permanently problematic character of teleology seems to be the most promising locus in his thought for a contemplative kind of philosophizing.

Teleological judging can be brought into nature, Kant argues, and only in a problematic way, by employing in the investigation of nature an analogy with causality in terms of purpose, whereby we think of nature as having a "technic."[41] Thus, we conceive of the concept of the object as its cause just as in human art, where the cause is the idea of the effect.[42] We can regard natural beings as having extrinsic purposiveness, insofar as they are useful to human beings; but this can be regarded as a natural purposiveness only under the condition that the existence of the beneficiary (the human species) is a purpose of nature in its own right.[43] This condition holds only with a particular understanding of the human (about which more later). Kant distinguishes this relation from the intrinsic purposiveness of natural beings, which we must understand as both causes and effects of themselves in three senses. (1) Members of a species produce other members of the same species, qua species both generating and being generated. (2) The individual produces itself in growth, which cannot be understood according to mechanical laws, for the individual does not operate simply on external matter. It develops itself by means of a material that is its own product, a matter peculiar to the species. (3) The part of the living being also produces itself in the sense that it contributes to sustaining the whole; there is a mutual dependence between the preservation of a part and that of other parts. Kant declares that these properties are "most marvelous" and "all our art finds itself infinitely outdistanced" by this mode of production.[44]

The situation can be expressed in terms of the traditional distinction between kinds of causes. Our understanding grasps causal connection in other beings only in terms of efficient causes that descend in a series, but our reason has the idea of final causes, as familiar to us in the practical sphere of artistic production, wherein an effect can be viewed as a cause in an ascending series. The dependence of the parts of a natural being on the whole we can conceive as production by a concept as in art, but the whole in the second case is not the cause of the product's matter. In the living being, the parts of the being (inclusive of matter) combine into the unity of the whole "because they are reciprocally cause and effect of their form." The productive role of parts can be seen as a connection of efficient causes and, at the same time, judged as causation through final causes. Each part is an instrument (organ) existing for the sake of the whole, but, unlike an instrument of art, the part produces other parts and the whole. The parts can thus compensate for damaged or deficient parts. A being with these characteristics is "both an organized and self-organizing being" and can be called a *natural purpose*.[45] An organized being is not a machine and can never be understood as one, for a machine has only motive force and cannot repair or reproduce itself. An organized being has formative force, about which "we say too little if we call this an analogue of art," a force inexplicable on any analogy with a known physical ability. The concept of a natural purpose is not "constitutive," as it does not express a known kind of causality.

Even so, we employ the remote analogy with art as a regulative idea to guide investigation and to meditate on the ultimate ground of these beings (*über ihren obersten Grund nachzudenken*)—a meditation that is not for the sake of knowledge but "for the sake of the same practical power of reason in analogy with which we considered the cause of that purposiveness" in organized beings.[46] This citation is an important indication of Kant's intent. The practical power in question is, in the first place, art; more widely, it is freedom, about which we reflect in connection with self-organizing beings. This accounts for some of the peculiar features of Kant's approach to living nature and teleology. Kant notably is focused on the productive aspect of living beings: reproduction, growth, nutrition, and self-production in general. His view of purposiveness does not address other natural ends or functions, such as locomotion, desire, perception, and judgment, as aspects of the organism, central to the account of soul as principle of life in Aristotle.[47] Kant's view of living beings covers plants and animals equally and abstracts from their differences.[48] The focus on the causality of self-production relates to Kant's effort to conceive of reason as autonomous in the radical sense of a self-organizing system. At the same time, his emphasis on the inscrutability of the power of self-organization prevents viewing this study as contributing, at least directly, to projects of mastering nature. The reflection on nature's formative power does not enable us to expand the human calculative-technical control of nature. Conversely, the reflection on our freedom provides limited insight into questions of how humans exercise causality over nature. I repeat my contention that, in one respect, Kant's thinking is an expansion of anthropocentrism, and, in another respect, it is a deep critique of anthropocentrism.

Culture as Nature's Ultimate Purpose

For Kant, that certain products of nature can be considered only in terms of final causes is a discovery, not only a subjective need.[49] The concept of purpose of nature is given objective reality by these beings, thus justifying the introduction of this concept into natural science, even though we have no a priori insight into the possibility of this causality.[50] The merely mechanical account of nature cannot explain organized beings, so that it is absurd (*ungereimt*) even to hope that perhaps, some day, another Newton might arise to show the laws for the production of a mere blade of grass.[51] The question of how such beings are possible is unavoidable; it is a permanent question for human reason.[52] All the same, we reflect on this question fruitfully. The experience of organized beings occasions the use of a maxim that has a priori status as inescapably necessary: that nothing in an organized being is gratuitous, purposeless, or to be attributed to a blind mechanism. This is implied by the formal definition of organized product of nature: it is a being in which everything is a purpose and reciprocally means.[53] Kant says we are entitled to go further with the reflection on purposiveness and consider the possibility that nature as a whole is a system of purposes. Within such a system, beauty in nature, as nature's harmony with the free play of our cognitive powers, can be considered an objective purposiveness of nature with respect to the human. For nature seems to hold the human in favor, by distributing not only useful things but also a wealth of beauty and charms.[54]

Yet such thoughts can be pursued only by leaving the account of phenomena and considering the possibility of a supersensible ground. The recourse to the supersensible is also needed to reconcile the mechanical and teleological modes of explanation, both of which are required in the study of nature as regulative principles. Kant says that although we do not know how far we can go with the mechanical kind of explanation, it must remain inadequate for things we recognize as natural purposes. The basis of reconciliation would be a mode of causation that is neither mechanical nor in terms of purposes, something we cannot grasp at all. It is possible that nature is based on an "intentional technic," in which mechanical laws are employed to achieve purposive organization.[55] This, of course, surpasses our power to apply mechanical explanations.

Thus, on the plane of speculative reflection, the ascent to which is justified by the experience of organisms, Kant considers the structure of the human as organized being and the place of the human species in the whole. That nature has a purposive structure beneficial to the human is a warranted assumption. The "idea" of purposiveness that provides a guiding thread for reflecting on human history in earlier writing is resumed in the third *Critique*.[56] In a note to the passage on beauty just cited, Kant asserts that, whereas in the judgment of taste no account is taken of the purpose of natural beauties, and whether they exist to arouse pleasure in us, here, in teleological judgment, "we may regard as a favor of nature that by means of the exhibition of so many beautiful shapes it would promote culture."[57] The critique of teleological judgment, and indeed the whole third *Critique*, could be said to culminate in Kant's account of how nature furthers human culture. He moves to that discussion via a distinction between ultimate purpose (*letzter Zweck*) and final purpose (*Endzweck*).

If nature is to be considered a system directed at furthering an ultimate purpose, something must exist in nature that can serve as the purpose. One can ask: for what end do the various kingdoms of nature exist? The answer is: "For the human being, for the diverse uses which his understanding teaches him to make of these creatures; man is the ultimate purpose of creation here on earth, because he is the only being who forms a concept of purposes for himself and who by means of his reason can make a system of ends out of an aggregate of purposively formed things."[58] But Kant proceeds to supply a condition on this human status: "It is man's vocation to be the ultimate purpose of nature, but always only conditionally: he must have the understanding and the will to give to nature and to himself a relation to a purpose that can be sufficient for itself independently of nature, which can thus be a final purpose, which, however, must not be sought within nature at all."[59]

Without man, the chain of mutually subordinated purposes in nature would not have a complete basis, for "only in man, and even in him only as moral subject, do we find unconditioned legislation regarding purposes." About the human as moral being, we cannot ask, "For what end (*quem in finem*) does it exist? His existence contains the highest purpose within it; and to this purpose he can subject all of nature so far as he is able."[60] It is the human supersensible character, the capacity to will an end beyond nature, which justifies regarding nature as having as *its* ultimate purpose the promotion of something "within man himself that is a purpose and that he is to further through his connection with nature."[61] As such, the human is properly titled the lord of nature (*betitelter Herr der Natur*).[62] It is "only as a moral being that man can be the final purpose of creation" and not as a being in pursuit of happiness, for "happiness is not even a purpose of nature with regard to human beings in preference to other creatures, much less a final purpose of creation."[63]

Kant offers some well-known comments on the problem of happiness. Happiness is a mere idea of imagination that cannot be made determinate and stable, so that even if nature were subjected wholly to the human will, no form of orderly satisfaction would result. The deepest difficulty lies not in the fact that nature has not adopted the human as its "special favorite" and exposes the species to as many external ills as other animals. It lies, rather, in the inability of human nature "to call a halt anywhere in possession and enjoyment and to be satisfied."[64] From this trait arise the peculiarly human miseries of oppression, luxury, and war. What remains as an end of nature, with respect to human nature, is culture, the aptitude for purposes generally, which nature can in various ways promote so as to prepare the human for its final purpose outside of nature. In the development of skill (*Geschicklichkeit*), the first form of this culture, nature employs the same seemingly unpurposive human traits to bring about results that humans do not intend. The natural violence of human relations makes necessary the lawful authority of civil society and, beyond that, war eventually propels competing states to form a cosmopolitan order of voluntary constraints on their external ambitions. Only in this order can the natural predispositions of humans develop maximally. "Though war is an unintentional effort of humans (aroused by unbridled passions), yet it is a deeply hidden and perhaps intentional effort of the supreme wisdom." At the same time, "War is one more incentive for developing to the highest degree all the talents that serve culture."[65]

The second form of culture is discipline (*Disciplin*), whereby nature "strives to give us an education that makes us receptive to purposes higher than those that nature itself can provide." By instituting the sciences and fine arts, nature refines taste and makes headway against the crudeness and vehemence of animal inclinations, promoting the sociable feelings and the communicable pleasures that make up humanity.[66] It is true that the same refinements produce the evils of luxurious ways and the use of knowledge for vanity, but, on the whole, they help liberate the human from the tyranny of the senses and prepare it for the sovereignty of reason. And again, even the evils play a positive role, insofar as the social corruption "calls forth, strengthens, and steels the soul's powers not to be subjected to those" evils.[67] The civilizing of human life is not itself moral but prepares for the moral.

The End(s)

The final end of Kant's philosophy, full autonomy of the human species, is one of entitlement but only ambiguously of accomplishment. By possession of a supersensible capacity, the human is "lord of nature," but nature in the relationship offers as much resistance as compliance. Nature is not an instrument for human happiness, but, even so, humans proceed with ambition, lust for power, and greed as though it were. The outcome is not what they seek but what nature intends, which is discipline and greater rectitude. Furthermore, the causality of nature as purposive is inscrutable and incalculable, and its assistance is an object of hope ("idea") and far from certain. Lordship over nature is not grounded in mathematical wizardry, which, in the realm of freedom, offers only illusion. Being lord of nature consists in good part in contemplating how little one knows; the abyss of ignorance is admiringly studied as an empty space of undefined possibility. In fact, one needs the prospect of unfinished striving, so that there is always something to will and always more hope. Clarity about how freedom relates to nature is, in principle, forever elusive, although reason must seek to realize freedom in the sensible world. The elusiveness is fortunate. Oddly, this gives Kant's philosophy, in the end, a contemplative aspect, one compatible with the seriousness of moral duty. One is urged to meditate (*nachdenken)* on nature's mystery, but the meditation is about interpreting nature as having an art that may, in a way unknown, remake the human. The reflection is sublime and ennobling, and seems to draw upon the thrill of not knowing whether one is acting or being acted upon (cf. the ambiguity of the feeling of the sublime as elevation and humiliation). Reason seems to have full control over itself as able to expose its own structure as a self-organizing system and thus overcome its metaphysical illusions. But as organic, its structure does not, in the end, allow transparency, which simply shows that the fundamental questions persist. Kant argues that philosophic architectonic is an art (*Kunst*) that partakes of the inspiration of genius; both the philosopher and the poetic genius, in his account, employ ideas whose original source and ultimate articulation remain obscure to the author.[68] And in both, nature creates as an artist, and art produces a new nature. One can say the final end of reason is its auto-production, in which one does not know, and cannot know, what the producer is.[69]

Chapter 12

Beyond the Island of Truth: Hegel and the Shipwreck of Science

Michael A. Gillespie

> This land is an island, and enclosed by nature itself within unalterable limits. It is the land of truth—enchanting name!—surrounded by a wide and stormy ocean, the native home of illusion, where many a fog-bank, many a swiftly melting iceberg give the deceptive appearance of farther shores, deluding the adventurous seafarer ever anew with empty hopes, and engaging him in enterprises which he can never abandon and yet is unable to carry to completion.
>
> —Kant, *Critique of Pure Reason*, A235/B294–A236/B295

The Problem of Science

We typically conceive of modern science as a tool that we employ to understand the inner workings of nature and to aid us in the development of a corresponding technology to improve human well-being. This, in any case, was the dream of Francis Bacon and his immediate successors. During the last hundred years, however, we have become more suspicious of these claims. In an obvious sense, we are less sanguine about the notion of scientific progress. The experience of World War I convinced many that science was not merely a progressive force that provides "new mercies" for human beings, as Bacon put it, but also new dangers, including ever more effective means of destruction. During the chaos of the interwar years and the horrors of World

War II, it also became clear that science not only facilitated the mastery of nature but the mastery of human beings, making possible new and more extreme forms of tyranny.

When facing the questions raised by these experiences, many still defended the neutrality of science, arguing that science and technology did not injure, enslave, or kill human beings; humans did. In attempting to defend this claim, they typically fell back on the distinction between facts and values that was given its most famous formulation by Max Weber. The source of this view of science, still popular today, was Immanuel Kant.

Kant's notion of science was part of his effort to overcome Humian skepticism that called into question not merely the possibility of science but also the possibility of morality and religion. In the case of science, Hume insisted that the belief in natural causality was misguided, since it imagined that the regularities it observed were naturally necessary, when, in fact, they were only habits of perception. Kant's transcendental idealism was an attempt to save science and make room for religion and morality. The success of this project depended on establishing a rational foundation for science and morality. The problem was that these two seemed to contradict one another, one rooted in the notion that every event has an antecedent cause and the other in the notion that individuals are only morally responsible if they are free. Kant saw this contradiction as an antinomy with a thesis, that there is causality only through nature, juxtaposed against an antithesis, that there is also a causality through freedom. In considering this antinomy, Kant argued that there could be no satisfactory causal explanation without assuming a first cause (drawing on Leibniz's principle of sufficient reason) and that a first cause would be a cause through freedom. But if there was a free first cause, then there could be a cause through freedom at any other point in the series, rendering all causal explanations untrustworthy. Kant then concluded that science and human freedom, the two pillars of the modern world, were apparently mutually necessary and mutually incompatible.

Kant imagined that he could resolve this antinomy through a critique of reason. He began with the assumption that everything we know, we know only through consciousness, and that all knowledge is thus subject to the limitations of consciousness. For Kant, the problem for reason arises when it exceeds its limits in an attempt to grasp the infinite. Science can only consistently explain the finite world. Science thus only gives us knowledge of what we perceive with our senses, that is, only the appearances of things, what he calls the phenomena. This is what is *for* us. In contrast to Hume,

Kant argues that we can thereby know not merely the concatenation of sense experiences but the *necessary* connections between them. Moreover, this is only one way in which we as conscious beings experience the world. We also experience the world through our awareness of the moral law; we thus know not merely what is but what ought to be. This insight, too, is rational but is not confined to or confirmed by sense experience. It provides us with access to the noumenal realm beyond the phenomena. In delimiting our capacities to their proper spheres, Kant believed he had resolved the Third Antinomy, since every event could have *both* a necessary natural cause *and* a cause through freedom.

Kant thus made it possible to imagine a natural science that did not need to concern itself with moral or other spiritual issues. Science need only explain the mechanisms of a universal efficient causality unconnected to divine or human ends. Science could bracket the actual world of human experience and imagine that another world, consisting merely of atoms or matter in motion, was the true reality that underlay everything and determined everything we do. In this way, the Kantian understanding of science provided the ground for a naturalism that has become increasingly indifferent and hostile to the very possibility of the existence of the realm of freedom and morality.

Kant, of course, did not go to such extremes. He believed he had saved *both* science *and* morality and shown that both were rational. The capacity to reason was both pure and practical, able to understand what is and what ought to be, each consistent within its sphere and at least potentially compatible with the other. This conclusion, however, was problematic, since it seemed to leave consciousness divided and incomprehensible. This bifurcation deeply concerned Kant's successors since, as Hegel put it, Kant saved the phenomena only by making human being itself contradictory.[1] Kant himself was aware of this problem but argued that this bifurcation was not fundamental but existed only within the transcendental unity of apperception, the unity of self-consciousness with itself. The character of this unity in Kant's thought, however, remained obscure.

In calling into question the comprehensibility of this notion of consciousness, Kant's idealist successors also called into question his understanding of science and morality. Kant depended upon drawing a sharp line between pure reason (and science) and practical reason (and morality). This line defined the limits of his famous "island of truth." However, Kant knew it would be extremely difficult to convince human beings to confine themselves to this island, especially since the island was constructed not from real things but only from appearances. It would thus be difficult to prevent humans from setting sail to explore the foggy, iceberg-infested seas.

This essay follows the voyage of G. W. F. Hegel on this foggy sea. I begin with an examination of Hegel's notion of a system of science and then focus on his philosophy of nature. I argue that his science is an answer to the problem Kant presents and is an effort to reconcile natural necessity and human freedom. I also explain why he thought it was necessary to reject an exclusively reductionist notion of science and to subordinate it within a system of science that incorporated elements of ancient and medieval science as well. Finally, I suggest that Hegel's voyage, for all of its daring, did not in fact take him far from Kant's island.

From Consciousness to Science

While almost all of the post-Kantian idealists accepted the Kantian premise that humans are essentially conscious beings, they all also sought to expand this notion of consciousness in a number of ways. For Kant, consciousness is in each individual, a reflective capacity for sensing and thinking that characterizes us as rational beings. For his successors, consciousness is something that is in us but also something that contains all of us and everything else as well. The social realm in this sense is not merely an intersubjective realm but a realm of general consciousness, or what Hegel calls spirit. Nature is not just a *res extensa*, as opposed to our *res cogitans*, but an integral moment of consciousness; and God is not an independent being but the absolute moment of consciousness itself.

The term for consciousness in German is *Bewusstsein*, which is a compound of *bewusst*, "known" (from *Wissen*, "knowing"), and *Sein*, "being." *Bewusstsein* thus might be literally translated as "known-being" or "being that knows." "Science" in German is *Wissenschaft*, and for Hegel it is the completion of *Bewusstsein* in the form of *absolutes Wissen*, "absolute knowing," that frees or absolves itself from *Sein*, "being." In freeing itself from its object, knowing becomes the pure idea that is recognized as the rational essence of nature and spirit, which includes individual, social, political, and cultural life.

Science, for Hegel, is the outcome of humanity's long and arduous voyage, a voyage of (self-)transformation and (self-)discovery, which he describes in a work originally entitled *The Experience of Consciousness* but that was afterward expanded and renamed *The Phenomenology of Spirit*. In this work, he recounts development of individual consciousness, then the development of general consciousness or spirit (social life), and, finally, the development of absolute consciousness (expressed in art, religion, and philosophy) to complete and perfect self-knowing or absolute knowledge. Absolute knowledge is, then, the foundation for science. The *Phenomenology* recounts the passage of consciousness and spirit through all of its possible forms, all of the different possible relationships between knowing and being, subject and object, driven by the desire of consciousness for reconciliation with itself and its inability to find satisfaction until its completion in absolute knowledge.[2] In each moment, knowing seeks to understand and become one with its object and, at the same time, to transform the object into itself. Every transformation of the object is thus also a transformation of the self, and vice versa. This was the famous dialectical process that Hegel believed came to an end in his own time and that made possible the speculative unity of consciousness (*Bewusstsein*); that is, knowing (*Wissen*) and being (*Sein*) in science (*Wissenschaft*).

At the end of this path of development, individual consciousness recognizes it can only be completely satisfied as a moment of the general consciousness of a rational community. General consciousness, or spirit, similarly recognizes that it is only satisfied as a community of autonomous rational citizens. And finally, the rational individual and community only come to properly understand themselves and their relationship to everything else when they realize that the absolute is not something above or apart from them but the very rationality that they have found in themselves and nature.

Absolute knowledge is the foundation for Hegel's system of science. This system has three parts. As pure thought divorced from being, it is an absolute logic of concepts and their relationship to one another, which Hegel describes in his *Science of Logic* and in the lesser "Logic" that constitutes the first part of his *Encyclopedia of the Philosophical Sciences*. These works provide an explanation of the categorical structure of reason that underlies everything else. This conceptual framework then makes it possible to understand the rational structure of unthinking being, that is, the rational structure of the natural world, which Hegel lays out in the second section of his *Encyclopedia*, "Natural Philosophy."[3] He then describes the reconciliation of these two in the completed science of individual, general, and absolute consciousness, or spirit, in the third section of the *Encyclopedia*, "The Philosophy of Spirit." As a system, Hegel's science is thus a coordinated, interlocking whole.

This system reconstitutes something like a traditional metaphysics. Metaphysics consisted of *metaphysica generalis* (ontology and logic) and *metaphysica specialis* (rational theology, rational cosmology, and rational anthropology). Kant's critique called this metaphysics deeply into question and was especially cogent in its attack on rational theology. Hegel accepts much of Kant's critique but then reconstructs metaphysics on a speculative or dialectical foundation.[4] His *Science of Logic* corresponds to *metaphysica generalis,* providing both an ontology and a logic.[5] His philosophy of nature and philosophy of spirit correspond to rational cosmology and rational anthropology, and underlying it all is his analogue to rational theology, the notion that reason is absolute.

While science, for Hegel, is the name for this system as a whole, I want to focus in the rest of this essay on Hegel's natural philosophy, in part because it most nearly corresponds to what we consider science and also because it is the most problematic section of his system.

Ancient natural science, beginning with the pre-Socratics, asked *what* nature was, in contrast to the efforts of earlier (mythological) thinkers who asked *who* the cosmos was. In asking this question, ancient philosophers, and Aristotle in particular, asked about nature's morphological character, that is, how it was divided up into different types of beings. Ancient science in this sense was, for the most part, ontologically realist, attributing real as opposed to merely nominal existence to universals and treating individual beings as mere exemplars of truly real species or genera. Insofar as it was realist, ancient science was also teleological, imagining that each individual being strove to actualize its potential by attaining its generic end, or telos. Ancient science was also principally theoretical, and while these scientists occasionally produced some practical results, like Archimedes they were embarrassed to have done so. The end of ancient science was thus not relieving human misery but participating in the eternal or divine.

Modern science took a different tack and asked not *what* nature was but *how* it worked. It was also preeminently practical, aimed at understanding how to alleviate the ills of the human condition and to make man master and possessor of nature. Moreover, the basic method of modern science was not morphological but analytical or reductionist, imagining that we can best approach the understanding of nature by dividing it up into the smallest possible parts, the ultimate particles or indivisible atoms, and then describing all other entities as the result of the motion and interaction of these particles by means of a universal mathematics (*mathesis universalis*). Universals or categories, from this perspective, are merely constructed entities that may be useful in everyday life but that are ultimately only fictions, in contrast to the truths of mathematics.[6]

In his philosophy of nature, Hegel seeks to incorporate elements of both ancient and modern science. His natural philosophy thus contains both realist and nominalist elements and is consequently, in many respects, at odds with the reductionism that forms the core of science today. This has led most modern scientists and scholars to dismiss his natural science. Already, in 1844, the biologist Schleiden described his philosophy of nature as a "string of pearls of the crudest empirical ignorance."[7] In 1915, Benedetto Croce assured his readers, in his famous *What is Living and What is Dead of the Philosophy of Hegel*, that the philosophy of nature was certainly dead. More recently, such eminent scholars as Charles Taylor and Robert Pippin, both interested in championing some aspects of Hegel's thought, similarly dismissed his account of nature.[8] What reason, then, do we have to reconsider it?

The need for a reconsideration of Hegel's natural philosophy is suggested by questions that have arisen concerning reductionist science in the later twentieth century and, particularly, its problematic impact on our notions of morality and human dignity. Hegel's goal of reconciling science and morality thus may offer us some solutions to our practical situation, but if his understanding of nature and science is defective, the force of these arguments will be blunted to say the least.

Many earlier objections to Hegel's account of nature have been answered in recent years. Already, in 1970, Michel Petry's annotated edition of the *Philosophy of Nature* made clear that Hegel's thinking about nature was not driven by apriorist or systemic needs, as many had long assumed, nor was it the product of ignorance. In fact, it was informed by some of the best empirical science of his day. This has produced a mini-Renaissance in studies dedicated to Hegel's natural philosophy that has helped to open up the questions I deal with in this essay.

Hegel, like many other thinkers before and after him, believed that while modern science offered useful explanations for how things work, it was not without its problems. He was convinced that it homogenized all things and thus did not attribute the proper metaphysical significance to different kinds of entities.[9] Matters of signal importance for human life were thus obliterated.

The livingness of living things disappeared in the process of vivisection characteristic of modern science.[10] Hegel also believed that the reliance on imagination and analogy were misleading, producing models not drawn from the world as it is but from some imaginary world, and further imagining that structures and relations that appeared in the macrocosm were equally characteristic of the microcosm.[11] And finally, he believed that the rough empiricism of much of experimental science made no sense without rational theories to explain both the axioms and results.

This said, Hegel was not an opponent of modern science. He accepted a great deal of what it had to offer, but he felt it was unable on its own to capture the richness of nature because it did not recognize the rationality of the structures that characterized experience. For Hegel, the world and everything in it is rational, or as he famously put it in the *Philosophy of Right*, "the rational is actual and the actual is rational."[12] Experience and the development of reason, for Hegel, are dialectical and speculative, and that means that while each stage is overcome, what is essential in it is preserved and sublated within a higher form of knowing. At the end of its development, general consciousness, or spirit, thus subsumes and embodies all of the previous forms of consciousness in absolute knowledge. Newtonian science thus is part of this final picture, but it does not provide a complete explanation of the whole. Hegel believed that even a superficial examination of Newton's thought made this clear, since he simply assumed, for example, that time, space, place, and motion were well-known and needed no explanation.[13] Fundamental elements of his thought were thus ungrounded. Moreover, he often used analogies that were never demonstrated or explained.[14] These concerns notwithstanding, in his account of mechanics, Hegel asserts that the Newtonian notion of gravity is foundational, but at the same time, he places mechanics within a larger and more differentiated vision of nature.

Perhaps most important for Hegel was the fact that, in eliminating all universals, reductionist science makes it impossible not merely to understand freedom but also the unique character and value of human beings. Humans, from this point of view, are essentially no different from stones, mere agglomerations of particles. Everything, in this sense, is homogenized, and consequently everything that we have previously attributed to particular categories of things, such as human rights and dignity, are simply illusions. Hegel considered the consequences of this view disastrous.

Hegel rejected this narrow vision of science and tried to lay out what he took to be the rational, ontological structures that characterized nature. It is possible, of course, to imagine these away, or at least to suspend them by imagining a world that God might have created, consisting of different entities and governed by different laws. Hegel, however, was convinced that we exist always only within the sphere of consciousness as the complex of knowing and being and that within this sphere, the rational is constrained by the phenomenological. Hegel thus conceived of nature as made up of phenomenological structures of increasing degrees of complexity.[15] His philosophy of nature begins not with matter but with concepts supplied by previous science and organizes them into a system according to the categories he detailed in the *Logic*.[16] Moreover, underpinning all of this is not a correspondence but a coherence theory of truth.[17]

At the core of Hegel's natural philosophy are thus natural laws and not natural entities.[18] This is significant, for if particles are primary, then the laws are merely a reflection of the particular nature of the particles and are not themselves primary or necessary. If laws are primary, then matter itself is simply an expression of the underlying rationality. Thus, when natural entities are replaced by natural laws, nature is idealized and rationalized.[19] In his philosophy of nature, Hegel thus argues that the rational principles are not imposed merely by the structures we use to understand nature but are actually in nature itself.[20] Indeed, this, for him, is what nature is. This fact helps Hegel overcome the problem of Kant's Third Antinomy.[21]

The goal of Hegel's science, as we noted above, was to overcome the bifurcation he saw within the Kantian notion of consciousness and, in this way, to reconcile freedom and natural necessity. Hegel believed that such a reconciliation was in fact the inevitable consequence of the historical experience and development of consciousness, which he had detailed in the *Phenomenology*. This reconciliation in fact defined the end of history. But how, then, does he reconcile freedom and natural necessity?

For Kant and Hegel, freedom means following the rules one makes for oneself, but the problem with this is that you first need a rule for making rules. Hegel solves this problem with his assumption that freedom is a social construct and achievement that is brought about by the dialectical development of general consciousness, or spirit.[22] The rules that we come to follow are rooted in our nature as *Bewusstsein* or consciousness, as the two-fold of knowing and being. The reconciliation of these two elements in our being is achieved in the rational state in which individuals live according to laws that they, as a community, give to themselves, and in the spirituality of art, religion, and philosophy. This vision of reconciliation, however, seems to leave out the impact of natural causality on the human body. Hegel's philosophy of nature is, in large part, intended to resolve this problem. Nature seems to us to limit freedom, because it is an alien other. The recognition that nature is fundamentally governed by the same rationality that governs us, however, makes it less alien and, at the same time, renders it more subject to human control. As rational beings, we thus come to understand ourselves as natural at the same time we come to understand nature as rational. This enables us to reconcile the two sides of our being. Nature becomes authoritative for us because of its rationality and thus is no longer an alien impediment to our will but an integral part of who we are and what we want to be.[23] It is thus possible for us to be free and autonomous beings in the world of Hegel's nature in a way that was impossible in the world defined by Newtonian science.

In summary, Hegel's system of science thus aims to demonstrate the unity of spirit and nature.[24] Both, in his view, are rooted in the idea, reason in its purest form.[25] What distinguishes them, according to Hegel, is that in nature, the idea appears as outside of itself; that is, as merely something in itself but not for itself, thus as exterior not only to spirit but broken apart within itself as separate atoms.[26] But it is fundamentally rational, and this is what we come to understand at the end of our process of development. As a result, nature no longer stands against us as an alien other but is recognized as an intrinsic element in our being as individuals, communities, and as the absolute.

Critique and Conclusions

The question remains whether Hegel's voyage was successful. I have dealt with the deep ontological problems of his system elsewhere and will not recapitulate that argument here.[27] Along with many others, I have also argued that his views about social and political life, as well as art, religion, and philosophy, are among the most insightful and profound in the history of Western thought. But for his system as a whole to succeed, his account of nature must also pass muster. And on that score, much remains mysterious and, in some cases, simply wrong.[28]

What is particularly perplexing about Hegel's philosophy of nature is its relationship to the rest of his system of science. The philosophy of nature stands crucially between the "Logic" and the "Philosophy of Spirit," but the transitions between the sections are extremely obscure. At the end of the greater *Logic*, Hegel suggests that the transition from the absolute idea to nature is a moment of liberation, in which the absolute idea sets itself free to become space and time.[29] Almost every scholar who has seriously studied Hegel's system of science struggles to make sense of this passage. J. N. Findlay, one of Hegel's strongest supporters, argues that "there is nothing but the utmost sobriety in Hegel's transition from the Idea to Nature," but he is unable to explain what it means.[30] The transition of nature to spirit is more familiar to us but still mysterious. Hegel moves from the inorganic world of physics and chemistry to geology, plants, and animals. Human animals then become conscious of themselves as generic beings and, thus, as spirit through the recognition of death. The role that death plays here is repeated on a number of occasions in Hegel's thought, most famously in the life and death struggle in the *Phenomenology*, but it remains mysterious on scientific grounds.

We can obtain some insight into these transitions if we remember that Hegel is trying to reconstitute something like a neo-Platonic Christian metaphysics. Hegel himself points us in this direction on a number of occasions. He describes the greater *Logic*, for example, as the "presentation of God as he is in his eternal essence before the creation of nature."[31] The question of how the general idea then determines itself *as* nature becomes, for him, equivalent to the question, "How did God come to create the world?"[32] This orientation is similarly apparent in the transition from the philosophy of nature to the philosophy of spirit. Within Christian neo-Platonism, the world is the creation of the God as the self-externalization of the divine essence as an other in order for God to come to see and thus know himself. Within creation, the crucial turning point is the coming into being of Christ, for that is the moment in which the world recognizes itself as God, in which the creature recognizes itself as creator. After Christ's death, this link to the divine is sustained and perfected by the Holy Spirit. From Hegel's systemic point of view, this is the core of the philosophy of spirit. His system of science is then complete when logic (God as the Word), the individual (the Son), and the community as a whole (Spirit) are recognized as an interlocking whole.

The problematic transitions in Hegel's system thus mirror the problematic moments in Christian neo-Platonism. Within Christianity, however, these transitions are achieved by the infinite and incomprehensible power of God. For Hegel, however, there can be no such miracles. His explanation must be thoroughly rational. Hence, while we may recognize the structures behind Hegel's account, his rationalist explanations of these transitions remain obscure and problematic.

In this respect, however, Hegel is no worse off than contemporary science, which assumes that there is a mathematical foundation to nature but has no explanation for why or how this is the case. Moreover, modern science is equally perplexed in its efforts to explain the notion of consciousness and the phenomenological world that consciousness opens up. To reject Hegel's science on these grounds in favor of reductionist science is thus unwarranted.

What is perhaps most troubling about Hegel's system of science, however, is that it remains deeply wedded to the Kantian notion that what is, is only in and through consciousness. Or to put it another way, Hegel may have attempted to sail away from Kant's island of truth but, in the end, and perhaps unbeknownst to Hegel himself, after his long voyage, he may have landed again on those very shores. Hegel's successors were certainly not convinced he had discovered a new world. Later thinkers, such as Nietzsche, came to believe that all idealists from Kant onward were Schleiermacher, that is, veil makers, who concealed the truth behind their conceptual illusions. Consciousness, in his view, is just the tip of the human iceberg, and what really matters is hidden beneath the surface of the psychic sea. If this is true, then understanding human beings and understanding nature will depend not on an exploration of what we can see and experience on the surface of the sea but what lies in the depths.

Kant believed we could live happily on his island of truth, comforted by "the starry heavens above and the moral law within."[33] Hegel, by contrast, was deeply troubled by the bifurcation this established in consciousness itself and searched for a way to reconcile consciousness with itself. His successors were not satisfied with his conclusions and set sail again, traveling further and deeper than Hegel had dared. Hegel sought to ensconce nature within consciousness; these sailor-thinkers sought to ensconce consciousness within nature. Hegel's voyage was less daring than theirs, but where they and we their fellow voyagers are traveling remains uncertain. Nietzsche pointed to the uncertainty and danger of this journey at the end of *The Dawn*: "Will perhaps someone say someday that steering toward the west we hoped to reach an India,—but that it was our fate to shipwreck on the Infinite? Or, my brothers, or—?"[34]

Chapter 13

Separating the Moral and Theological Prejudices and Taking Hold of Human Evolution

Lise van Boxel

To see a being as Nietzsche sees it is to see its history, its genealogy. It follows from this that philosophy is genealogy. As an inroad to this genealogical thinking, we will use Nietzsche's presentation of his biographical development as a philosopher. This philosophical autobiography is found in the "Preface" to his book, *On the Genealogy of Morals.*

Nietzsche begins by introducing himself. He tells us he may almost be defined by what he calls "a scruple," peculiar to him, concerning morality. By his own account, it entered his life "so early, so uninvited, so irresistibly, so much in conflict with . . . [his] environment, age, precedents, and descent that . . . [he] might almost have the right to call it . . . [his] '*a priori*'" (*GM*, "Preface," 2).[1] This scruple soon took the form of a question: "What is the origin of good and evil?" (*GM*, "Preface," 2).

He goes on to report that, at the ripe age of thirteen, he dedicated his "first philosophical effort" to this question. In the essay that was the fruit of this effort, he gave the "honor" of having generated evil to God, a conclusion he judges "was only fair" (*GM*, "Preface," 3). The older Nietzsche may be delighted by his younger self's answer. Nevertheless, he is proud to report that his development as a philosopher advanced because he rejected this response. He explains, cryptically, that this rejection followed from the fact that he "learned early to separate theological prejudice from moral prejudice" (*GM*, "Preface," 3). Precisely what are these prejudices?

Nietzsche speaks frequently in his writings of the moral prejudice. It is the conviction that good and evil are universal values. The good is allegedly good for all human beings, and evil is purportedly wicked for everyone. What Nietzsche means by the theological prejudice is less clear. Since Nietzsche tells us the moral and theological prejudices are distinct, the theological prejudice cannot denote the belief that God is the supreme good, for this prejudice resolves into the moral prejudice. Rather, the theological prejudice must refer to the belief in an eternal, unchanging, and absolute or pure—in other words, unmixed—God. This concept of God is essentially the same as the traditional philosophic notion of eternal, unchanging being, which is absolute or pure because it is wholly unmixed with becoming (see *BGE* "Preface").[2] Such being does not exist in the world in which we live, for our world is characterized by transfiguration and becoming. To maintain the truth of this concept of God or being, therefore, one must posit an otherworldly realm that exists behind our world and that transcends becoming.

As opposites, absolute being and becoming do not exist on a continuum with each other. They are not different by degree. Rather, they are wholly different in kind. Thus, the theological prejudice is an expression of the larger category of prejudice—namely, the "*faith in opposite values*," which Nietzsche identifies as the "fundamental faith of the metaphysicians" (*BGE* 2). This faith results from a conflation of the theological and moral prejudices, which produces the notion that eternal, unchanging, and absolute being is universally good. Almost all philosophers to date have believed in this compound prejudice: "It looms in the background of all their logical procedures; it is on account of this 'faith' that they trouble themselves about 'knowledge,' about something that is finally baptized solemnly as 'the truth.' . . . It has not even occurred to the most cautious among them that one might have a doubt right here at the threshold where it was surely most necessary—even if they vowed to themselves ['all is to be doubted']" (*BGE* 2).

Nietzsche then invites us to reflect on each of these prejudices. Regarding the moral prejudice, he asks whether attributes characterized as evil are as valuable to the vitality of the human species as the so-called good attributes. He then recombines the moral and the theological prejudices and asks us to wonder whether the value of the alleged absolute, universal good consists precisely in the fact that it is not the *opposite* of evil but is instead *one* with it: "For all the value that the true, the truthful, the selfless may deserve, it would be still possible that a higher and more fundamental value for life might have to be ascribed to deception, selfishness, and lust. It might even be possible that what constitutes the value of these good and revered things is precisely that they are insidiously related, tied to, and involved with these wicked, seemingly opposite things—maybe even one with them in essence. Maybe!" (*BGE* 2; see also *GS* 4).[3]

Return to Nietzsche's claim that the separation of the two prejudices was crucial to his development as a philosopher. Recall, he tells us that, before he learned to separate these prejudices, he attributed the origin of evil to God. What we did not note previously, but what is significant, is that he says nothing of his early conclusion about the origin of the good. Initially, one might think he does not speak of it because the young Nietzsche casually attributes its origin to God, the supreme good. Hence, its source is not a question for him. However, this inference does not adequately explain why the mature Nietzsche, who does not attribute the good to God, nevertheless chooses to speak only of his encounter with the question of the origin of evil. Given that his immediate subject matter is his philosophic growth, it is reasonable to infer he thinks the question of the origin of evil in particular highlights the hindrance to philosophic development that results from conflating the moral and theological prejudices. More specifically, reflection on the origin of evil illuminates the path to a new understanding of morality.

The young Nietzsche initially presumes God created all things. Thus, when confronted with the question of the origin of evil, he concludes it, like everything else, was generated by God. Given this, good and evil have a common source. According to the theological prejudice, however, this is impossible. Opposites cannot have a common source, since their opposition is absolute. Thus, what appeared to be an answer to the question of the origin of evil raises more vexing questions, questions that demand some difficult decisions.

If one wants to retain the theological prejudice, one must conclude evil did not originate in anything purely good. For example, evil cannot result from a confusion or dissipation of goodness. Rather, since God is presumed to be the epitome of pure goodness, evil must be generated by a power that is wholly independent of, and absolutely other to, God. Thus, there must in effect be two creators: one purely good and one purely evil. On what grounds are these creators characterized as good and evil? To make comparisons and contrasts, one must slide back and forth along a continuum of similarity and difference that precludes absolute otherness or opposites, and one must have a satisfactory sense of how far into otherness one has moved. In all such movements, one is necessarily adulterated.

Since purely good and purely evil deities cannot intermingle, their moral characterizations must depend on a standard separate from either of them. This standard raises two problems for the hypothesis. First, the necessary recourse to an independent standard means neither the good god nor the evil god is the true authority in moral matters: the authority is the standard itself. Second, by positing the existence of a standard that incorporates both moral values, one effectively returns good and evil to the same source. One has thereby reconstituted the god to whom the young Nietzsche attributed the origin of both good and evil. *This* god is no longer pure. Hence, it cannot be the god of the theological prejudice.

One might try again to salvage the theological prejudice by claiming God is both absolutely good and omniscient. One would have to admit human beings cannot understand how both of these things can be true, but one might assert they are both true nonetheless. Someone might add that it is precisely because we cannot understand such essential truths that we must accept God's authority.

The crux of this hypothesis is that the human perspective is insurmountably limited and false, insofar as it is mixed up with becoming. This claim attracts Nietzsche's attention. By thinking through its implications, he realizes moral valuations must ultimately articulate human, not divine, judgments.

The premise of this hypothesis implicitly recognizes that our experiences must be filtered through what we are as human beings. Speaking more precisely, they must be filtered through what we are as individuals. These observations do not entail the additional conclusion that either the human perspective or the individual perspective is absolutely and universally true. On the contrary, from within our perspective, we can discern that other beings do not have the same perspective as we. Like our own, their perspectives are integrated into, and generally support, their physio-psychologies, which are more or less like ours (see *GM* 3.12; *GS* 374).

Dogs clearly experience smells more acutely and vividly than we do. The hummingbird, which moves very rapidly compared to the human being, must have a different sense of speed than humans. What appears to be extremely rapid motion to us must be normal to this bird. If the hummingbird's sense of speed were *not* calibrated to its physio-psychology, but were instead the same as ours, it would be inviable. Similarly, the mayfly, which in its mature form lives for twenty hours, must experience time in a manner calibrated to its physio-psychology in order to complete the activities that constitute a typical lifecycle for its species. If a mayfly could consider an average human lifetime, it might believe this time is unfathomably long—an eternity. If it could reflect on what it might be like to live so long, it might say, "Life is made meaningful by our consciousness of death! If life were significantly longer or—God forbid!—eternal, I would be bored to death! Surely, eternal life is valueless!" Buzz, buzz.

We can try to bridge the gap between non-human and human perspectives. For example, we can use technology to increase our sense of smell in order to gain some sense of a dog's capacity. Even in this case, however, our heightened smell must be experienced by our non-dog selves. The human being will never have the same experience as the dog so long as the two types of beings are distinct.[4]

Our inability to transcend our human perspective in our relations with the other beings in the earthly realm is equally applicable to a world that might exist behind our own and to the other-worldly beings that might populate it. Even if one were to receive commandments directly from a supposedly transcendent realm, the human perspective remains unavoidable and insurmountable for us, since we must still understand the revealed commandment. Understanding requires interpretation, which necessarily entails making judgments about meaning. Having made these preliminary judgments, one must then deliberate about whether to obey the command as one has understood it. All of these judgments are human. Nietzsche follows this line of reasoning to its conclusion. Since we cannot know anything supra-human, the supra-human simply does not exist for us. With this conclusion, he frees himself from the theological prejudice, according to which there is a true world of eternal, unchanging, and absolute being behind our world (see *D* 130; *TI*, "How the 'Real World' at Last Became a Myth"; *GS* 346).[5]

Overcoming the theological prejudice has a decisive consequence for the moral prejudice. People's belief in the universal authority of the dominant moral values depends on their belief in eternal, unchanging, and absolute being or God. In other words, the moral prejudice depends on the authority of the theological prejudice. With the destruction of the theological prejudice, therefore, the foundation for the authority of the moral prejudice is also overturned (see *GM* 3.24).[6]

This realization might prompt you to try immediately to discover a new foundation for the authority of an absolute, universal good. Why? What is your motive? Whatever the answers to these questions may be, and they may be different for different people, the effort is premature. At this point, we ought to realize we do not even know adequately what morality is, let alone whether it is worth preserving (or whether it can be preserved). We have spiraled back to the young Nietzsche's questions regarding morality, only now the inquiry is on a higher plane than the one on which he, and we, began, for the moral-theological prejudice has now been cast aside. We must survey this new territory to determine what the guiding question of morality has become.

Since everything that exists for us must be interpreted and judged by what we are, valuations are necessarily always human: consciously or unconsciously, we create them. Thus, an investigation of morality and valuation is necessarily an investigation of the human being. Having realized this, we become more interested in knowing the creator—the human being—than the creations. The creations are valuable first and foremost as a means to understanding the creator. Broadly speaking, valuations articulate something about the physio-psychology of the human being or of certain human beings, but what precisely is articulated? Notions of good and evil or good and bad. Yes, of course, but what do these values *signify?* Of what are they *manifestations?* In sum, the question concerning morality now becomes: what is the value of morality *for human life?*

Brief as his account is, Nietzsche gives us enough insight into the path he took at this juncture of his investigation for us to see we are on the same track as he. He tells us that, after he ceased looking for the origin of evil behind the world, his question regarding morality was transformed. He, too, began to ask what the value of morality is in terms of human life. He fleshes out the meaning of this question by articulating related questions that illuminate it:

> A certain amount of historical and philological schooling, together with an inborn fastidiousness of taste in respect to psychological questions in general, soon transformed my problem into another one: under what conditions did man devise these value judgments good and evil? *And what value do they themselves possess?* Have they hitherto hindered or furthered human prosperity? Are they signs of distress, of impoverishment, of the degeneration of life? Or is there revealed in them, on the contrary, the

> plentitude, force, and will of life, its courage, certainty, future? (*GM*, "Preface," 3; see also "Preface," 5)

Having identified the new form of the question concerning morality, Nietzsche reports he "discovered and ventured divers answers" to it, answers that vary according to historical epochs, peoples, and "degrees of rank among individuals" (*GM*, "Preface," 3).

If we now draw back from the details of the image Nietzsche sketches of his own growth and consider the picture as a whole, we can see that it shows us something of what it is to be a living being, and of what genealogy is. A living being has a certain degree of organized, self-directed, and continuous motion, which is affected by things extrinsic to it. The history of this motion and its current condition merge fluidly in a continuum that determines various possibilities for its immanent motion, growth, or decay. This fluid motion—the motion of the becoming that is the being's being—is what Nietzsche means by generation. Thus, genealogy is the study of the changing shape—the morphology—of life.

We can now also see why it is fitting that Nietzsche introduces genealogy with a description and a metaphor rather than a definition—such as, genealogy is the logos (λόγος) of generation or growth (γενεά). By so doing, he provokes us to *envision* this particular kind of motion, whereas, if he had immediately offered a definition of the word, we would be inclined, at least initially, to overlook the kind of becoming that is a living being and instead to mistake this being for something static. In other words, a definition would reinforce the error that has plagued almost every thinker to date. While this error is due to the conflation of the two prejudices, it is also bound up with words, which of course are themselves static. We can see continuous motion, but we cannot speak of it directly. While imagery employs words and therefore cannot entirely overcome the stasis they involve, it also provokes us to use our mind's eye to see the motion that is described. Thus, the image communicates more effectively the becoming that genealogy addresses.[7]

These clarifications help us better understand Nietzsche's question of what the value of morality is for human life (*GM*, "Preface," 3). With this question, he is asking what particular moral doctrines reveal about the vital condition of the human beings who create them and who believe them: "Are they signs of distress, of impoverishment, of the degeneration of life? Or is there revealed in them, on the contrary, the plentitude, force, and will of life, its courage, certainty, future?" (*GM*, "Preface," 3). In addition, he is asking whether and how specific moral doctrines help the human being, or particular human beings, secure the fundamental good—that is, the experience of their maximum vitality: "Have they hitherto hindered or furthered human flourishing?" (*GM*, "Preface," 3). In sum, moral doctrines are valuable to the extent that they promote our growth. From the standpoint of human evolution, moral doctrines—which human beings create—have so far been the most momentous means to our experience of maximum vitality, for they have been one of the most significant factors, if not *the* most significant, in determining what we have become.

Now, see how Nietzsche's description of his own growth illustrates this answer to the question of the value of morality for human life. His moral scruple is the commanding thought that compels him to pursue the question of morality to previously unknown depths of the human physio-psychology. As a result of his explorations, he discovers a new way of understanding the world, which prompts him to reject key aspects of the moral-philosophic tradition as threats to his otherwise ascending vitality. In lieu of them, Nietzsche aims to live according to his understanding of the *fundamental* good. In keeping with his understanding, he realizes his philosophic inquiries are good because they develop his soul into a more comprehensive, vital whole. That Nietzsche intends his depiction of his own growth to illuminate the value of morality for human life in general is indicated most clearly by his placement of it at the beginning of a book on the genealogy of morality, which proves to be inextricably intertwined with the evolution of the human being.

In spite of the fact that most people, including most thinkers and scientists, now accept evolution as a true account of living beings, they continue to cling to the notion of a final cause or natural limit that determines what a living being is, all that it could ever be, and what it ought to be. However, the notion that a living being is subject to an insurmountable limit that defines both what it is and what is good for it is an expression of the moral-theological prejudice (see *BGE* 9, 22; *GS* 37, 109, 143). There are no such limits, nor is there any reason to think they would be good, if they did exist. The absence of such limits is what it means to be a being that is a becoming. In other words, traditional definitions of the human being define what we are ahistorically, but these definitions are inadequate because a living being is a type of motion that is inseparable from its genealogy. Nothing with a history can be defined according to a single, ahistorical moment in its genealogy, as though this moment encapsulated all that it ever was or could become (see *GM* 2.13).

In lieu of this traditional approach to understanding a being, Nietzsche's account of his own development indicates that, if one wants to know what a being with a history is, one must know its genealogy—one must know the narrative of its changing shape as a living form. If one wanted to know something about a living being with a view to determining whether it is evolving or declining, or with a view to guiding its motion, one might reasonably look to the activity that is most vitalizing for it, or that has the potential to be most vitalizing. One might seek to augment this activity or to make it more of what it is. Nietzsche does this in his own case when he speaks of his moral scruple. He does it again in the case of the human being. As a result of his genealogical investigations of the species, he concludes that the human being's most vitalizing activity to date consists in our ongoing effort to take charge of our own evolution. So far, we have engaged in this project of self-determined evolution unconsciously, instinctually, and largely by means of moral doctrines (see *GM* 2.1–2.2).

Where might such taking charge of our own evolution lead? While our genealogy determines to some extent what we are, the boundaries it imposes on us are not absolute. Given sufficient time, any aspect of what we are could recede or disappear from our physio-psychology, if not from our history:

> That the character is unalterable is not in the strict sense true; this favorite proposition means rather no more than that, during the brief lifetime of a human being, the effective motives are unable to scratch deeply enough to erase the imprinted script of many millennia. If one imagines a human being of eighty thousand years, however, one would have in him a character totally alterable: so that an abundance of different individuals would evolve out him one after the other. The brevity of human life misleads us to many erroneous assertions regarding the qualities of man. (*HH* 1.41)[8]

Thus, we come to realize the living being is neither absolutely fluid, nor is it fixed in its present state. It has a past that informs, and to some extent determines, what it is in the present and what it can become in the immediate future. Pathways of development that were available to it at one stage of its development are not available to it in the same way later, after its physio-psychology has been altered. Regarding its future, there is no limit on what a being might become. For example, the current physio-psychology of the human being strongly inclines us to suspect we cannot acquire the capacity to regenerate all parts of ourselves in the next instant or tomorrow. Nor do we expect suddenly to acquire the capacity to see electromagnetic fields. We regard these conclusions as reasonable because we have not discerned a path to such immediate and dramatic alterations in our physio-psychology. However, the fact that such capacities seem unavailable as the immediate next step in our evolution does not rule out the possibility that the species will acquire them at some future stage of its genealogy. Similarly, while it may be true that every human being that has existed to date has died, this does not mean mortality is necessary to what we are. That a being lives does not necessitate that it also dies.

Chapter 14

Mastery of Nature and Its Limits: The Question of Heidegger

Mark Blitz

Is Modern Technology Novel?

Mastery of nature has come to mean overcoming or relaxing natural limits. The natural limits one usually has in mind are the physical or bodily limits of human beings, limits of communicating, traveling, seeing, and dying. To overcome natural limits means to reduce or eliminate long times, vast distances, debilitating diseases and suffering, and to increase life spans. One might ask why overcoming or attempting to overcome such limits should concern us.

What makes these limits of speed, distance, sight, and health natural? One reason is that they are our animal limits. We cannot go far on our own or communicate with people far away. We are subject to illness and to our unaided life span. But, we also talk and think. Why is our freedom and reason less natural than are our bodily restrictions? If reason is natural, why are bodily limits more reasonable than overcoming them? We are always "mastering" previous restrictions. Moreover, the power of speech and reason itself seems to expand (or diminish) through accumulated understanding. So, it seems that there is nothing novel about our mastery of nature or, therefore, about modern technology. Mastering or overcoming seemingly inevitable limits is what we always do.

Let me press this opinion, because, by contrast, many, including Heidegger, believe that modern technology is decisively new and often problematic. The usual candidate for this novelty is mathematical physics. Everything is at root the same, from the dominating viewpoint of measuring movement and place; what differs—however things look—are only the numbers and interactions. Because everything is this way, the more powerful our knowledge becomes, the more we can make things behave as we want.

On this view, ancient art and modern technology do differ significantly. Ancient art helps things to become what they are as they follow along their natural path: farming or "cultivation" is the obvious example. Or, it sets things back on their natural path: medicine is the obvious example. Or, when ancient arts do produce new things—houses or clothes—they still follow the capabilities of visible natural materials: wood, stone, skins, and metals.[1] But, by contrast, modern technology need not follow visible paths or materials. So, computers and telephones are brand new things whose production does not mold visible natural materials or take a natural process and help to ensure its completion.

One counter to this view of what is radically new in modern technology is that while some contemporary technological implements seem altogether new, most only enhance, harness, or copy a natural object or process: birds' flight, natural magnification, miniaturization in mirrored reflection, echoes, lightning, and so on. Modern technology powerfully understands how things happen, but anything it makes, and how it makes it, follows things and processes that are already there. At the end of the day, contemporary technology is only an unusually powerful new tool.

There are several replies to this counter. One, as we said, is that some of what today's technology creates is so novel as to seem brand new. Not airplanes that look like natural objects, or even pharmaceuticals that regularize production or delivery of what already exists in natural plants and other substances, but computers with no obvious natural analogues are what are emblematic. Of course, computer processes do have a kind of natural analogue or example: humans. So, perhaps the novel element is the way we now do (some) things that are grounded in violating our commonsense understanding of causality, space, and time, and, as such, make sense only mathematically.

A second reply to the view that there is nothing essentially novel in modern technology is that the pervasiveness of technology is so vast that even if most of its products and processes have natural analogues, when we take these products and processes together, we have something that is effectively altogether new. The diminishing importance of local differences, instant communication, new ways to effect or limit conception, and lengthening life have so changed the unity, tempo, and attachments of everyday existence that matters have changed radically.

A third reply is that while the usual arts and modern technology have changed many things about our bodies, our mastery of nature is now such that we are about to transform our minds, and therefore ourselves, radically. Within a century, we will be a half-different species, or the old species with our rank at the top replaced.

A fourth reply to the opinion that modern mastery of nature is not radically new is Heidegger's reply. His view is simple, but it is difficult to grasp. Because he supported the Nazis, moreover, one is afraid that to agree with anything he says sets one on the path to perdition. Heidegger's thought is linked to environmentalism, furthermore, and extreme environmentalism is also politically suspect, in sound quarters. So, Heidegger's worries about the mastery of nature are difficult to grasp, and if one agrees with them, one is in bad political company. Why, then, make the effort?

Heidegger's Understanding of Technology

To understand Heidegger's simple view of technology, we need to say something about his understanding generally. Every entity with which we deal first comes to light within a meaning or intelligibility of a certain sort, which he calls its being. To use a hammer as a hammer, for example, and not to see it as a block of material, we must first understand implicitly that the hammer, or any tool, belongs to an order of functions and involvements: this here, used for that, for the sake of this. We always find ourselves in some world of meaning and involvements: even what is absent or missing is meaningfully absent. We are never isolated, unmoved subjects who gaze at a completely incomprehensible outside. One reason for our immersion in significant worlds is that a human being is the entity whose own being is always an issue for him, a being that I determine, for a while. My determining a possibility of my being—that for the sake of which I am, at the time—is central to forming orders of significance and to letting things emerge in their own being as, say, ready-to-hand tools. We are always open to our own being (because we always need to determine it) and to the being of what we are not (because we are always immersed in contexts that involve other entities). We are thus always open to being simply.

This openness and immersion is not occasional or random: we must understand each of our human characteristics in terms of our openness to our own being and to being as such—our speech and understanding, our moods and dispositions, our meeting of our needs. Our characteristics, therefore, are not defined by our being rational animals, calculating animals, desiring animals, freely willing entities, or any of the other usual alternatives. Our openness to or understanding of being is indispensable for the meaningful presence of anything, including ourselves. Everything that has meaning must first reveal itself in an open field before we secure it in any way, or precipitate particular entities in a meaning, and we ourselves are first of all by extending the dimensions of and standing within this field. We almost always forget or ignore this responsibility, but we sometimes face it, in dealing resolutely with moods such as anxiety.

To be free is to stand within this openness. This openness is "historical," moreover, because it and we are finite—we are our being only as some possibility on which we decide for a while. Our openness is not eternal. What we can never master, then, is just this: that all our characteristics belong to our openness to being, and that there is no meaning or being without us. What we also cannot master is that we always find ourselves in our characteristics (or our being) in some particular way and are always thrown into some world. We can never fully master ourselves or being because we cannot go fully behind or ahead of them. Because our chief characteristic is that we are open to being (or meaning, or presence), and this being is not an entity that is caused (because causality has no meaning without us, and we are always by understanding our possibilities and are thus never fully actualized), we are essentially beyond the causal nexus of entities, even those that are highest and most general.

Heidegger's view of technology is that it is an understanding of being, and not primarily a set of implements. Technology as a dispensation of being is the ground or context for "technology," understood as the products of mathematical physics; it does not equal such products. The meaning of technology is that everything is first seen or understood as a possible resource for a continuing uncovering of resources. Technology is the framework within which everything is challenged to stand ready to approach us as an endless reserve. We can then grasp these resources in the identical mathematical, temporal, and spatial terms that allow us to manipulate matters endlessly, but to no result in particular. Nature must show itself to be capable of being mastered and ready to be mastered before it can be mastered. This showing and approach is what technology is.

We can clarify what Heidegger has in mind in still another way. To speak concerning nature's mastery requires us to understand nature, mastery, and their link. I say "nature's mastery" to retain and focus ambiguity about whether the issue is nature's mastery over us or ours over it. It is precisely such ambiguities that belong to our comportment to being. This relationship indicates the possibility that the link between mastery and nature is not an occasional connection into which each half may or may not enter, but an essential connection on which we draw implicitly when we discuss each part separately. The meaning of man and of nature in technology is linked. So, we must also, or especially, consider the "of" or "over" when we discuss mastery of nature.

Attention to the link between mastery and nature, and to the coexistence or ambiguity of the subjective and objective genitive, characterizes Heidegger's thought generally and is among the foci he usefully and distinctively develops. Man's contemporary mastery of nature belongs together with nature's contemporary mastery of man. We today, in Heidegger's view, conceive or approach both man and nature as technological entities. Everything is energy to be unlocked. Everything loses its distinctive form and end. Everything, including man, is first seen as a resource for the continual production of resources. Everything is challenged and challenges itself to come forward and to be grasped in this way. Technology, in the sense of scientifically constructed implements, and even mathematical physics itself, is consequent to technology as a frame of challenge and resource.[2]

Such technological understanding allows nature to be or seem to be "mastered," in that we can direct its forces to yield products and effects that differ from what occurs prior to our intervention. It is this understanding, however, rather than any new products, that is the heart of technology. Such mastery, moreover, is inseparable from technologically understanding man himself. It is not the case that through technology we have new powers but that we otherwise remain the same. Rather, everything, including us, loses its independence and becomes merely a cog in a machine. Man is intelligible today as technological man, a mere fragment. Still, our technological understanding and self-understanding exemplify the truth that in all we see and do, we have a prior understanding of being, and that this openness is the unmasterable ground of human characteristics.

Nature and Technology

The dominance of technology is both an historical change and a continuation. It is a change because it is connected to a novel understanding of man as a subject and of what he understands as objects that oppose him, and that he knows by representing them to his self-certainty. We associate this change with Descartes and his attempt to find secure, certain knowledge rooted in what we cannot doubt. Although Descartes and Cartesianism are not inevitable, they do not appear from nowhere. They belong to our understanding of being, and to the epochs of philosophical or metaphysical understanding.

To consider previous epochs of being is to see that nature was not always associated with technology.[3] Heidegger's view is that nature, from the pre-Socratics at least through Aristotle, means the emergence of things on their own, as their own, within their own limits. At first, nature meant this emergence, or things in their emergence. The emphasis was on the emerging, not the thing. But although this was the emphasis, the pre-Socratics, and surely Socrates, did not attend to the emerging as such, or to the context that allows emergence to occur—truth, the open, temporality, presence—or to how this context is given. In Plato's thought, nature still names what emerges on its own, but the emphasis shifts to the idea, that is, to the fully emerged entity or thing once one sees it intellectually and as it seems to be, eternal and unchanging. What is natural about a thing is what the thing is, not that it is; its permanent form, not its fleeting embodiment: the thing's being is its whatness.[4]

Seeing what is as what is natural relies on taking for granted the meaning of being as presence and, therefore, on being as what is most present. This involves being eternal but, as we indicated, it also suggests that a natural thing is what stands fully within its limits, as beginning from its limits, as it were. In this sense, what counts as being takes its lead from production as well as presence: in Aristotle, one can understand limits most clearly from what is produced or produced completely. This view of being as the natural is connected to being as what is most readily available, as one's household property and possessions are available. So, what is most available—emerges on its own, is most present, and stands completely in its limits—is what emphatically is, and this is the natural.

Nature understood in this manner could not be mastered because what is natural cannot be broken down and reassembled without destroying it. What we can say, however, is that how to combine natural things in different materials or among themselves is not altogether set in advance. One sees this politically, for example, in the differences among regimes and cities. Heidegger himself does not discuss this question.

How Heidegger Deals with Technology: I

Heidegger's radical understanding of technology is for him the clue to dealing with it properly. To first have things approach us technologically is to understand everything, including ourselves, in terms of technological being—the human being as human resource, social capital, instrument of production, and so on. But technology is not the only understanding of being, as we just indicated. There are various "historical" ways in which things first approach us meaningfully, from Plato and before to Nietzsche and after; technology is closely connected to will to power. These dispensations of being also include seeing things as tools in their interconnected functionality, or merely as present. In any event, our basic human characteristics are ultimately to be understood as what allows us to stand out and come back toward being in its several historical dispensations and modes. We stand out and come back by understanding our possibilities, and by letting things be so that they can affect us, as a mood, such as fear, reveals things as fearsome. For various reasons, we fail to grasp ourselves this way: we interpret our characteristics in terms of the way we deal with and see non-human entities, and we grasp these entities (and, therefore, ourselves) explicitly in terms of the prevailing metaphysics. In our contemporary case, we see things so narrowly, that is, technologically, that explicit understanding has lost touch even with philosophy, so that the "truth" of things is only their concrete scientific characteristics. Any sense that things must first come to light as possible or meaningful in certain ways, let alone that the central human characteristics are only properly understood in terms of the way we are co-responsible for this revealing, is almost completely forgotten.

Yet, this almost complete occlusion of being and *Dasein* (man as open to being) by technology allows us (or Heidegger) to notice what is missing: the being and man always overlooked but always there, and always in some way experienced. In this sense, the dominance of technology, its driving out of other ways of experiencing and interpreting, allows an echo of the larger context that allows it to be. The mastery of nature finally permits us to see what we cannot master, namely, the fact that human beings are always open to being. What gives meaning or presence to entities as technological, or what allows them to present themselves technologically, is neither restricted, reduced to, nor predictable by this single meaning; what defines humanity is the way we are open to being. Of course, humans once were not and need not be always.

How Heidegger Deals with Technology: II

To develop the argument, we should say more about our openness to being. Heidegger differs from previous thinkers because he connects this understanding to our particular, concrete situations, not to our being equivalent instances of a generality. I am historical and individual because my being is always an issue for me and my possibilities are always projected forward from the world and the public into which I find myself thrown. My being is mine to determine concretely, but I usually see myself and my possibilities as others do, as the "they-self" embedded in a public world. If and when I do understand myself authentically, as my own, I grasp my unity by projecting my dying (the possible impossibility of all my possibilities), by resolutely taking over my thrownness, and in readiness for moods such as anxiety. When I am authentic, I also reveal my public authentically, as a "people," and illuminate my tradition as the authentic heritage of my generation.[5] Heidegger's emphasis on the particular, individual, historical, and one's own is not the usual historicism, however, because he does not reduce us to an effect of a causal nexus, whether of class, age, or material causality. My connection to a people and its possibilities and to my own self belongs to my openness to being, presence, or meaning. Particularity and individuality, peoples and heritages, living within and projecting the dimensions of temporality, are necessary for there to be any intelligibility at all. No other can take *Dasein*'s place. In more traditional terms, which, as such, Heidegger would reject, no intelligible presence or meaning exists apart from the "this," this particular, the imperfect.

What, then, is Heidegger's "solution" to the technological covering of man? His first and always present effort is to bring forward his novel understanding of man and being and, therefore, of technology's limits and dependence. Exactly where a proper, or resolute, understanding of being might lead, however, is unpredictable. Heidegger also means to deal with technology by reminding us of possibilities for dealing with things meaningfully that, because of technology, we now neglect: everyday use and the contexts that make this possible; how simple things—jugs, dwellings, the items of daily life—that stand on their own are; what we take for granted in living with these things; art and, especially, poetry and its source; genuine education.[6] He also always retains a hesitation or antipathy toward regimes that he believes to be technologically reductionist in his sense, American liberal democracy no less than Soviet Marxism.[7]

This antipathy reminds us of Heidegger's Nazism, which is the second way that he tried to deal with technology and its effects, hoping that this regime was rooted in an authentic German people.[8] Heidegger was rector of his university in Freiburg during parts of 1933 and 1934, defended Hitler's coordination or amalgamation of activities (thus closing off their independence), and taught things in seminars from 1933 to 1935 (for which we have Heidegger's notes and student reports or protocols that he approved) that support German expansion and the elimination of Jews, at least from Germany, if not our destruction simply. Heidegger's politics cause one to be extremely cautious concerning the need to "do" something about technology.

The Limits to Technology

Perhaps, however, we can employ Heidegger's understanding to deepen other approaches to understanding and directing the mastery of nature. Here, we begin in an obvious way by asking what modern technology is for. We usually treat it as a means that we may control or direct, at least in principle. From Heidegger's standpoint, this is already a mistake because it presumes that our contemporary ends exist, have meaning, or can be formulated apart from technology. As Heidegger sees it, however, everything, including the presentation to us of the meaning and presence of the "goals" to which we would lead technology and the force or structure by which we would direct it, has already been experienced technologically, that is, as we have said, through a structure or framework in which everything is challenged to appear as and is broken down into a formless resource for endless continuation.[9] Even apart from Heidegger, indeed, we can observe the link between modern technology and a larger liberal democratic context of liberating acquisition, freeing human energy, overcoming human abasement and excessive attention to an afterlife, and ending mastery of ourselves by others.

We can suggest, in contrast to Heidegger, that while there is indeed a link between liberal democracy and technology, the driving force of technology is not a new understanding of being, but a changed conception of the substance and manner of completion of human ends. While this conception differs from past views, moreover, it does not differ from it radically in its meaning. Acquisition and the virtues connected to it belong to a view of the human good as satisfaction of desire, such that overcoming penury, disease, war, and, where necessary, time and place are desirable or necessary. This view, indeed, goes together with particular ways in which goods are experienced in relation to their unity, wholeness, and partiality: goods conceived and experienced as, in principle, equal objects of desire characterize modern thought. But, they are not experienced wholly or only in this way. The goods sought through modern technology—health, wealth, comfort, and so on—retain much of their traditional or natural meaning, attraction, and intelligibility. The modern revolt against otherworldly humility and oppression is a revolt against the dominance of "piety." But this revolt and the goods it seeks are not so fully reconceptualized, and human enjoyment so newly experienced, that these goods, traditionally and liberally understood, cannot remain an independent standard by which to judge contemporary technology and liberal democracy. In fact, one source of the appeal of modern technology is that it has meaningful effects for good or ill in the everyday world, which indicates not only its power but its connection to a more general understanding.

From this point of view of the question of what technology is for, we can also discern limits that technological mastery can never overcome. One set of goods that no change, technological or otherwise, can simply master are those in which exclusivity belongs to—cannot fully be separated from—the good experienced or, indeed, where exclusivity constitutes a central element in its desirability. A prime example is loyalty or truthfulness: being true. One cannot be equally faithful to more than one or attentive fully to more than one. Love, integrity, faithfulness: truthfulness. (Heidegger sees this in his way in his considerations of ownness and resolve.) Why is this? It is partially because of the shortage of time, but more time and space cannot change this restriction because it involves the relation of particular to particular in their particularity or identity. One can consider here what Aristotle means by the friendship of the virtuous. There is a limit to mutual open attentiveness: the meaning (or goodness) of the good restricts how much can be produced and enjoyed, where. At some point, of course, this necessary limit can become remote from the ordinary loyalty of love and family and life and death, but the limit cannot simply disappear. And, as we know, there are goods, such as honor and rule, which diminish if shared or change in their enjoyment if they become too widespread: exclusivity belongs to what makes them desirable. Moreover, and connecting again to our first point, no technology can eliminate the pleasure or virtue of doing things on one's own, employing one's own efforts or virtues. An outcome can be assured, but the path itself and its pleasures cannot be replaced, but could at best be simulated. But perhaps here and elsewhere the illusion could replace the reality? It would be an illusion nonetheless, and one therefore would have the good in a truncated way. This disjunction between the real and the illusory is perhaps most visibly and powerfully evident in seeking to understand. To be truly enjoyed, knowledge must be known as such, and the knower must know himself as such. Technology cannot overcome this limit: illusion contradicts knowing in any experience of it. In all these instances, it is not only what is true or good but our own experience, openness to, and enjoyment of what is true and good that sets unbridgeable limits.

The limits to enjoying and understanding such goods are, in my judgment, intelligible beyond or before what Heidegger believes to be our current technological world. Perhaps this is because these goods, once understood, reflect something eternal and unchanging—natural forms, as Plato might suggest—so that every human openness to or experience of them must finally be similar. As Heidegger sees it, however, to orient ourselves to "eternal" things is to orient ourselves ultimately to causes outside ourselves and, therefore, to begin to lose our free grasp of our uniquely human experiences. In his argument, our distinctive freedom and unity are never constituted by a free-floating reasoned direction to something unchanging but, rather, by the understanding, thrownness, and standing in significance that I discussed earlier. Everything eternal and simply unchanging must first present itself within what is meaningful and, hence, temporal, finite, and historical—a structure that allows a meaningful future, past, and present. Even technology's inability to overcome mathematical or other immaterial truths would rest finally on its inability to overcome our being, our responsibility for intelligibility.[10] Our grasp of the immense range of things that we can show to be does not come from glimpsing the eternal. Rather, anything that is and, indeed, being or meaning as such, must join with us in our finitude to have its strength and power. Technology, for Heidegger, is the last stage in forgetting our finite openness and seeing ourselves only as entities in a causal chain that leads nowhere. Human understanding and freedom, however, always come to light in what is not yet, in open possibility in which even inevitable dying and thrownness, the negative, must be experienced. This element of the negative and not yet is inherent in everything and cannot be overcome technologically. Man can be eliminated, but the limits to our being cannot be, and, if Heidegger is correct, no entity without the experience of anxiety, reverent silence, perplexity, or wonder—without the possibility of the end of possibilities, without the experience of thrownness or history that limits our possibilities, and without the hermeneutic circularity in understanding that is attendant upon meaning and self-direction—could understand being or its possible range, for being is co-respondent to *Dasein*. Being cannot fully be, cannot be brought out in its fullness or developed, without us.

Limiting Heidegger

Let me turn briefly to my final topic, the question of whether we can ground human and political guidance differently from Heidegger while retaining his insights. If he is correct that we are not subjects looking out at objects, how else may we capture our finitude? If we are not merely reasoners inferior to computers and machines, what defines us?

Technology occurs as and within a context of intelligibility that outstrips any concrete entity, technological or otherwise. This outstripping involves possibility, which outruns concrete actualities but always belongs to them too. If possibility cannot be itself except as it dwells together with us, moreover, it is chained to finitude, or to limit, however far one ascends. Another way to express this is to say that the meaning (presence, intelligibility, direction, possibility) as which everything presents itself also contains what is not so, in terms of this meaning.

Is Heidegger's the only way to grasp this link between intelligibility and finite limit?[11] If we consider, say, Plato's dialogues, we see that everything presented as whole or as beautifully independent is incomplete in fact. Every regime falls short of justice. Each idea is elusive. Any whole in which the ends or goods reside that are crucial for ordered meaning is imperfect. The ambiguities of politics and of opinions perhaps even suggest that we can freely understand only if understanding is incomplete, for how else could things be allowed to stand out for us in a full range of their complex connections and separate powers?

Somewhat more concretely, one may say that problems, perplexities, and wonders, the ridiculous, honorable, and beautiful—all the phenomena that show our individual and political limits—are inherent in the intelligibility of things, but not only in Heidegger's way. The unity from which we actually begin, which we may then split into "that" and "what," is not the authentic individual and his heritage, but the imperfect yet striving unity visible in laughter, perplexity, loyalty, opinion, and wonder—that is, through a "self-consciousness" that spirals upward and not only stretches out and back. Such an imperfect unity, individually and politically, allows prudent guidance, because it is less trapped within itself than Heidegger's *Dasein* and his people.[12] Of course, to contrast such a view with Heidegger convincingly, one would need both to describe it in detail and show how such limits allow "history," or variety in meaningful new philosophies, arts, mathematical physics, and ordinary modes of being.

Chapter 15

What Is Natural Philosophy? The Perspective of Contemporary Science

Adam Schulman

The common premise of this book is, I take it, that when modern natural science was founded about four hundred years ago, it made a decisive break with ancient and medieval science, embracing the mastery of nature as its overarching goal in place of a merely contemplative knowledge of nature. The question I pose here is whether contemporary natural science, despite this turn toward mastery, is still capable of yielding not only fruit but also light. Modern science has undoubtedly given us powerful tools for predicting and controlling the phenomena of nature. Does it offer, in addition, profound insights into the intelligible nature of things? Does it provide genuine knowledge of nature that substantially meets the requirements set by Aristotle in his *Physics*, where he says that such knowledge must disclose the *causes*, *principles*, and *elements* that are responsible for the being and becoming of each thing? That is, despite the methodological turn that modern science took at its founding—despite its sharp break with ancient speculative natural philosophy and its preference for power over knowledge, or at least power with knowledge, is it possible that *precisely* modern science—and above all *modern theoretical physics*—has fulfilled, at least in part, the speculative program of ancient natural philosophy?

* * *

But first we must consider whether, for the ancients, a science of nature—a deep, comprehensive, precise understanding of the natural world—was a realistic goal of philosophy, an impossible dream, or somewhere in between. Every serious student of Plato and Aristotle must, at some point, have lost sleep wrestling with this question, if only because of the strikingly different answers these two philosophers seemed to give to it. Plato, following his teacher Socrates, *seems* to have given up on the possibility of a natural science, or at least to have turned away from the direct investigation of nature to the dialectical examination of men's opinions about the human things. Aristotle, for his part, invested immense time and energy composing numerous treatises on various topics in natural philosophy, as if a science of nature were very much a serious goal of his philosophic inquiries.

Plato has Socrates say in the *Phaedo* that in his youth he was wonderfully desirous of that wisdom called "inquiry concerning nature." It seemed to him "magnificent to know the causes of each thing, through what each comes into being, and through what it perishes, and through what it is" (96a5–7). But Socrates concluded that he had no natural aptitude for this kind of investigation, for it so blinded him that he unlearned even what he and others thought he clearly knew before (96c3–7). Eventually, he tells us, he turned away from the investigation of "beings" and took refuge in "speeches," embarking on his "second sailing" (99d1–e5). Aristotle reports, in the *Metaphysics*, that Socrates "concerned himself with ethical matters, and not at all with nature as a whole" (987b1–2) and, in the *Parts of Animals*, that "with Socrates the inquiry concerning nature ceased, and those who philosophized turned toward useful virtue and politics" (642a28–31).[1]

As for Aristotle, one finds among his extant works treatises on physics, generation and corruption, the heavens, meteorology, the soul, and the history, parts, motion, and generation of animals, among many lesser works on natural philosophy. Well over half the pages of the Corpus Aristotelicum are devoted to the inquiry concerning nature. In fact, logic, metaphysics, and natural philosophy together comprise about 1,100 of the 1,500 pages of Immanuel Bekker's standard edition, leaving fewer than four hundred pages for "the human things"—ethics, politics, rhetoric, and poetics. Aristotle gives every appearance of taking natural philosophy quite seriously, and of believing that a science of nature is sufficiently available to us to justify expending enormous efforts inquiring, thinking, and writing about every aspect of the natural world.

How should we understand this apparent discord between Plato and Aristotle on the question of the availability of a science of nature? One interesting and, perhaps, plausible answer is to be found in David Bolotin's 1998 study, *An Approach to Aristotle's Physics*.[2] Bolotin's extremely careful reading of the *Physics* attempts to distinguish between that work's rhetorical surface and its philosophic core. It turns out, on Bolotin's reading, that most of what have come down to us as Aristotle's doctrines in natural philosophy were invented by him for ulterior, esoteric purposes and were not meant seriously as principles of a genuine science of nature. By the end of Bolotin's study, an impressive collection of venerable Aristotelian doctrines in natural philosophy lie discarded on a growing heap of merely rhetorical, exoteric teachings. These include the doctrine that in every natural motion or change there is a persisting underlying substrate, and that natural beings consist of a form that is present *in* such a substrate; that ends or final causes have a role to play in natural motions; that natural magnitudes are continuous, in the sense of being infinitely divisible; and that, more generally, the reach of natural science need not be limited to what is perceptible to human beings but can render intelligible what lies beneath or beyond the perceptible. Perhaps most significantly, Bolotin claims to show that the manifest central teaching of Aristotelian physics—that there are principles, causes, and elements that are responsible for the being and the coming-into-being of natural things, and that these principles, causes, and elements are both intelligible in themselves and, in some measure, knowable by human beings—is entirely false, and that the true Aristotelian teaching is that *there are no such principles* (Bolotin 22). Thus, according to Bolotin, Aristotle comprehensively and deliberately misrepresented his views on the availability of a science of nature, pretending that the natural world is far more intelligible than he actually believed it to be. As for Aristotle's purpose in perpetrating this elaborate deception, it was mainly, according to Bolotin, to defend the reputation of the philosophic inquiry into nature and to protect from persecution those who undertake that inquiry.

If Bolotin is correct in his assessment of how seriously we should take at least the surface teaching of Aristotle's works on natural philosophy, then the perplexity as to the apparent disagreement between Plato and Aristotle on the availability of a natural science is largely resolved. That is, they both turn out to have agreed with their master Socrates that natural science is, on the whole, unattainable—that nature is, at least for human beings, hardly knowable at all. There remains, of course, the question of why so many Bekker pages of Aristotle's writings are devoted to *apparently* serious inquiries into nature, but that can perhaps be explained by the severity of the danger in which philosophy found itself in the ancient world and the corresponding extremity of the need for defensive rhetoric. Be that as it may, Bolotin ends his study with the suggestion that, while his surviving works on natural philosophy present, at least on the surface, a largely false or at least wildly exaggerated picture of Aristotle's view of the intelligibility of the natural world, "the chief way in which he tries to lead us toward his own perspective on the world is through his *political* philosophy, which we find above all in the *Nicomachean Ethics* and the *Politics*" (Bolotin 153, emphasis in original).[3]

Setting aside the problem of classical natural philosophy, there remains the question of natural science as it has developed in the 2,300 years since Aristotle and, above all, the question of the modern scientific project begun by Galileo, Bacon, and Descartes in the 1600s and continuing to the present day. Must there not remain a powerful suspicion that, in the four centuries in which the modern scientific investigation of nature has flourished—and, in flourishing, utterly transformed the world we live in—we have accumulated evidence pointing to a far more favorable assessment of the intelligibility of nature?

Bolotin, for his part, does not think so. He quotes Albert Einstein himself (in a 1938 book he co-wrote with Leopold Infeld called *The Evolution of Physics*[4]) as writing that "physical concepts are free creations of the human mind, and are not, however it may seem, uniquely determined by the external world." The laws of nature discovered by modern science thus have, as Bolotin asserts in his own name, "no evident connection to the ultimate causes of the natural world." They carry "no conviction except insofar as they help to predict observable events" (Bolotin 33). According to Einstein and Infeld, "We can well imagine that another system, based on different assumptions, might work just as well." The natural scientist, they write, is like a man trying to understand the mechanism of a closed watch: "He may form some picture of a mechanism which could be responsible for all the things he observes, but he may never be quite sure his picture is the only one which could explain his observations. He will never be able to compare his picture with the real mechanism and he cannot even imagine the possibility or the meaning of such a comparison" (quoted in Bolotin 33–34). In short, modern science only builds more or less plausible and convenient models for the purpose of predicting (and, we might add, controlling) observable events, and it never claims to truly understand the reality underlying those observations. Such models are put forth not as true accounts of the intelligible nature of things, but merely as useful tools for saving the appearances.

Coupled with Bolotin's own interpretation of Aristotle's *Physics*, this judgment as to the modest theoretical scope of modern science would seem to mean that the conclusion reached long ago by Socrates—that a science of nature is simply not available to man—remains valid, undisturbed not only by Aristotle's many writings in natural philosophy, but even by the four hundred years in which the modern scientific project has flourished. Whether and to what degree this claim has merit is not, it seems to me, a question that can be decided by armchair philosophers. It would require some considerable effort to understand the principal theoretical achievements of modern science, with a view to deciding whether they are in fact merely heuristic models for predicting observations or something more like the intelligible principles, causes, and elements that Aristotle, in his *Physics*, says are the hallmark of a genuine science of nature.

Setting aside whether this extremely modest account of the theoretical reach of modern science is generally accurate (or even whether it accurately captures in particular the theoretical contributions of Albert Einstein), one may ask what might be the ultimate source of that alleged modesty, and here I finally turn to the subject of this book, the mastery of nature, understood as the core aim that the modern scientific project assigned to itself from its inception. For it may plausibly be argued that modern science as a whole retains to this day the essential character given to it by its founders four hundred years ago, the chief feature of which is a fundamental turn away from speculative inquiry into the intelligible nature of things and a reorientation toward practical mastery of nature in the service of human freedom, comfort, and security. I will not rehearse that argument in detail here, other than to point to the theme of "methodology" in Descartes, Bacon, Hobbes, and others among the founders of modern science; the universal method adopted by modern science was intended to guarantee the reliability and utility of its results rather than the theoretical truth or comprehensiveness of the picture of nature that emerged. Nature as a whole is not assumed to be inherently knowable by man; but recalcitrant nature may be dragged piecemeal into the laboratory and vexed until she reveals some of her secrets, thereby enabling us both to predict and control an ever-increasing fraction of the phenomena of the world we live in.

I am not questioning the claim that natural science took such a methodological turn at the end of the sixteenth century, a turn that has largely reoriented the goal of modern science toward the conquest of nature, nor do I doubt that—in the service of that goal—the modern project has methodically narrowed the scope of what counts as "knowable" for the purposes of science—embracing a dogmatic materialism, for example, not because it is self-evidently true that the world consists of nothing but bodies, but because it may prove useful to assume so. But I would like here to re-open the question of whether modern science, despite those self-imposed limitations, might still, in the course of four hundred years of experimental and theoretical inquiry, have made profound discoveries that even the Ancients would have to admit shed a genuine, if partial, light on the intelligible nature of things. I want to focus in particular on certain insights provided by the two great achievements of twentieth-century theoretical physics: relativity and quantum mechanics.

Let me concede at once, in limited agreement with Einstein and Infeld, that modern physics never makes the boastful claim to have achieved the ultimate and definitive understanding of any domain of nature, or to be confident either that no experimental evidence will ever appear that contradicts the current theory, or that no superior theory will ever be proposed. Let me also concede that what I am claiming as insights into the intelligible nature of things do not in any sense amount to a comprehensive picture of nature as a whole, making sense of all its phenomena, both human and non-human—far from it. But that lofty and perhaps unattainable goal need not be satisfied in order to make the case that contemporary theoretical physics constitutes at least a partial realization of a science of nature as envisioned by Aristotle.

Aristotle says at the beginning of the *Physics* that "we think we know each thing when we know its first causes, and its first principles, and [have followed it back] as far as its elements" (*Physics* 184a12–14). I think Aristotle is deliberately vague here as to what he means by "causes, principles, and elements" (and even as to the meaning of "first"), and there is in fact considerable overlap in their ranges of meaning. To know the principle (*arche*) of a thing, we learn from *Metaphysics, Book Delta* (1012b34–1013a23), means to know in some way where it comes from, its origin, source, or foundation, what makes it what it is, or what makes it intelligible. "Cause" (*aitia*, 1013b24–1014a25) covers a similar range of meanings as "principle" but includes specifically what something is made of, its form or shape, what sets it in motion, and what (if anything) it is aiming at. Broadly speaking, the cause of anything is whatever answers the question "why?" about that thing. Why is it the way it is, why does it do what it does? Elements (*stoicheia*, 1014a26–b16), on the other hand, would seem to mean primarily the fundamental building-blocks out of which a thing is composed and into which it can be analyzed, much the way a word is made up of letters.

My suggestion in what follows will be that modern physics has made discoveries that by any reasonable standard amount to causes, principles, and elements in the Aristotelian sense of those terms. In my opinion, formed over some years of studying the history of physics, and especially the momentous developments in theoretical physics in the early twentieth century, the modest, merely heuristic view of the scope of modern science that David Bolotin finds support for in the book by Einstein and Infeld is just *too* modest. To make this suggestion more plausible, and perhaps to whet the appetite for further study of the relevant theoretical physics, I sketch here two examples, one a principle (*arche*) discovered by the theory of relativity, the other an element (*stoicheion*) revealed by quantum mechanics.

Absolute Space-Time

Einstein's special theory of relativity, published in 1905,[5] has altered forever our understanding of space and time. Until Einstein, it did not occur to anyone that the spatial separation between two different events—how far apart they occur—and their temporal separation—how much time elapses between them—have any interdependence on one another. Thanks not only to Einstein but also to Hermann Minkowski, who a few years later gave a beautiful and definitive four-dimensional mathematical framework to Einstein's theory,[6] we now *know*—as confidently as we know anything, I would argue—that there are no such things as space and time; there is only "space-time." We now know that the space-time interval between any two events has an absolute magnitude, irrespective of any frame of reference from which we measure it. And that four-dimensional magnitude is measured using a strange new "metric" that differs in a few decisive ways from the Pythagorean theorem that governs distances in Euclidean space. Meanwhile, what we used to think of as the space separating two events, and the time that elapses between them, now turn out to have no absolute magnitude or even existence on their own but merely a relative or perspectival existence and magnitude. In effect, space and time are mere projections or, as Minkowski put it, shadows cast on the particular axes of our chosen frame of reference by the absolute interval, whose magnitude is itself measured by the strange pseudo-Euclidean metric of four-dimensional space-time. Reflecting on this insight, Minkowski was led to affirm that "the word 'Relativity-Postulate' for [Einstein's theory of space and time] seems to me very feeble; . . . I prefer to call it the 'Postulate of the Absolute World.'"[7]

I would say that Minkowski's discovery (made possible, of course, by Einstein) of the absolute existence of space-time, and of the non-existence of a separate space and time, constitutes a *first principle* of physics in the Aristotelian sense. That discovery, which has not yet been challenged in any serious way, despite its shockingly strange character and the passage of more than a century, has had profound and far-reaching implications for every branch of fundamental physics. We now know that there is no such thing as an electric field or a magnetic field; there is only an electromagnetic field. There is also no such thing as the energy or the momentum of a given object; there is only a four-dimensional thing, called the energy-momentum four-vector, whose absolute length turns out, astonishingly, to be the object's *mass*. The invariant length of the energy-momentum four-vector is connected with yet another remarkable insight of special relativity, namely, the equivalence of mass and energy. We now know that every object of mass m, when looked upon in its "proper" frame of reference, in which it is at rest, has an intrinsic energy content equal to mc^2, where c is the universal speed of light in empty space.

Furthermore, we know now that there are two fundamental species of space-time interval, which Minkowski named "time-like" and "space-like." Only a time-like interval is capable of coinciding with the "worldline" or four-dimensional path that an object describes as it travels through space-time. If an object is traveling in "inertial motion," maintaining a constant speed in a definite direction, it is always possible to transform one's frame of reference to the object's proper (i.e., resting) frame. In that frame, the object's trajectory through the four-dimensional world consists entirely of time; the length of any space-time interval in its history is made up only of time and not of space, and we call time as measured in this privileged frame of reference the object's "proper time." It follows that the truest measure of experience or history is not the space an object traverses but the time that elapses in its proper frame.

Finally, only two years after introducing the theory of special relativity to the world in 1905, Einstein, while seeking to extend the scope of that theory, was struck by a thought that he later recalled as "the happiest thought of my life."[8] What occurred to him was that for a body in free fall, *there is no gravitational force*. Someone who has jumped off the roof of a house or out of an airplane simply *does not feel or experience his own weight*. We are therefore free—and indeed obligated—to deny the existence of that force and to consider a body in free fall due to gravity (whether a person falling toward the ground or, say, a planet orbiting the sun) to be experiencing not motion accelerated by an external force but *inertial* motion; in fact, one may think of such a body as being *at rest* in its proper frame throughout its entire history. This happy thought led Einstein, eight years later, to the formulation of his general theory of relativity, which is both a generalization of special relativity (to include arbitrary reference frames) and, at the same time, a disclosure of the uniquely geometric nature of gravitational attraction: that gravity is not, strictly speaking, a force at all but rather a change in the local curvature of space-time wherever and whenever energy-momentum is present.

There can be no doubt that the theory of relativity and the discovery of space-time have made significant contributions to the modern project of the conquest of nature. For example, Einstein's famous equation $E = mc^2$ played a certain role in the development of nuclear fission and the testing of the atom bomb, though not the central role that is often mistakenly attributed to it. There is also no doubt that the theory of relativity was made possible by the extraordinary achievements of many great scientists, such as Newton, Faraday, Maxwell, and Lorentz, who themselves were working within the project of modern science inaugurated by Galileo, Descartes, Bacon, and others. But despite the entanglement of Einstein's achievement with the modern project dedicated to the conquest of nature, it seems to me perverse to deny that what Einstein and Minkowski achieved and gave to us was a *profound theoretical insight into the nature of things*—things that, as it turns out, we did not really understand before, including space, time, energy, momentum, mass, and gravity—an insight that has never been challenged in any serious way and that, it seems likely, will remain a fundamental principle of physics for as long as human beings inquire into the nature of things. In my opinion, the postulate of absolute space-time is therefore fully entitled to be called a *prote arche*, or first principle, of nature in the Aristotelian sense.

The Quantum of Action

My second example, drawn from quantum physics, concerns not only principles (*archai*) in the Aristotelian sense but also elements (*stoicheia*). In pursuing the nature of things "as far as the elements," one might reasonably suppose that what one is after are the fundamental irreducible units of which the physical magnitudes of our world are composed; and one would hope that the phenomena of nature would somehow disclose those basic units to us in the course of experience and, if need be, experiment.

For example, is matter composed of ultimate irreducible atoms of some determinate mass, as various atomists have suspected? Is energy composed of ultimate irreducible bits of energy? It is fascinating to learn that nature does in fact disclose to us a certain universal minute quantity of "stuff" of which everything in the world is (in a sense) composed—but the unit in question is not one of energy or mass but of *action*, a term coined in the eighteenth century by Leibniz, equal to energy multiplied by time, or momentum multiplied by distance. The Leibnizian action is not an instantaneous quantity like energy or momentum but a quantity associated with an entire physical process from start to finish—for example, the trajectory of a particle from one space-time point to another. As we shall see, the fundamental unit or quantum of action is not an indivisible *atom* of action, but a universal *measure* by which the action in any process is to be counted and interpreted.

The route to the discovery of the quantum of action is a complex one, but the main points are as follows. First, in 1900, Planck was able to explain the unexpected distribution of electromagnetic radiation emitted by a "black body" on the assumption that harmonic oscillators in the walls of the body are able to vibrate only at a series of *discrete* energy levels—classically, a completely unexpected and unjustifiable assumption. In dropping from one energy level to the next lower one, such an oscillator, vibrating at frequency v, will emit radiant energy in a tiny *discrete amount* (quantum) of frequency v and energy $E=hv$, where h is "Planck's constant," to which Planck pragmatically assigned a magnitude that made his theory fit the observed energy distribution of black body radiation. For our purposes, it is noteworthy that since E is in units of energy and v is in units of 1/time, Planck's constant has units of *action*, and when considered as a quantity of action, its empirically determined value is ridiculously small: about 6.6×10^{-34} joule-sec.

Next, in 1905, Einstein was able to explain the strange pattern of emission of electrons by a metal surface when light shines on it, on the assumption that light of frequency v strikes the metal not as a continuous wave spread out in space but as a collection of discrete, localized packets (*light quanta*, later named "photons"), each with energy $E=hv$.

Then, in 1913, Niels Bohr was able to explain the strange discrete spectrum of light emitted by atoms as their electrons fall to lower orbits, by assuming that only certain, discrete electron orbits are allowed, the frequency v of the emitted light being determined by the difference in energy $\Delta E = h\nu$ between the higher and lower orbits. Bohr postulated that the allowable electron orbits are those on which the electron's *angular momentum* is $nh/2\pi$, where n is a whole number—as though *angular momentum* always comes in quantized *units* of size $h/2\pi$.

Then, in 1915–16, W. Wilson and Arnold Sommerfeld independently gave the Bohr postulate a new interpretation: the allowable orbits are those for which one full cycle around the orbit has an accumulated *action* equal to nh—as though *action*, at least in periodic processes, comes in quantized units of size h. Wilson and Sommerfeld generalized this statement into the postulate that all periodic phenomena in nature (orbits, vibrations) obey the *action quantization condition*

$$\oint p\,dq = nh,$$

where p is the momentum, q the corresponding coordinate, and the symbol $\oint$ indicates integration around one full cycle of the periodic process.

Finally, in 1922–1924, Prince Louis de Broglie, starting with the Einstein relation $E=h\nu$ between the energy and frequency of a photon, postulated that *every material* particle of mass m, and hence rest energy mc^2, is "associated" in its proper ("resting") frame with a *periodic phenomenon*—a sort of internal clock—of frequency $\nu = mc^2/h$. Purely relativistic considerations led Broglie to realize that, when the particle is moving with momentum p and energy E, the postulated internal "clock" would give rise to a *traveling phase wave* of frequency $\nu = E/h$ and wavelength $\lambda = h/p$.

De Broglie's postulated internal clock in effect gives every body of mass m a way of counting the passage of time in its own proper frame. For if the frequency of that clock (in cycles per second) is $\nu = mc^2/h$, then the period of the clock (in seconds per cycle) is $T = 1/\nu = h/mc^2$. Imagining that the internal clock has a traditional face with a single sweeping hand, that hand will complete a full rotation every T seconds. At any given time, the *phase* of the clock, corresponding to the *direction* in which the sweeping hand is pointing (0 at 12 o'clock, .5 at 6 o'clock, etc.), is equal to the proper time t divided by the period of the internal clock T, or in other words, the number of full rotations the clock's hand has made.

The phase $t/T = tmc^2/h$ of the postulated internal clock also has an interesting relation to the action. For if the action along a particular path through space-time is equal to the time-integral of the energy along that path, then for a particle at rest in its proper frame, the action will be simply the internal energy of the particle times the elapsed proper time. When divided by Planck's unit of action h, this is a dimensionless number tmc^2/h, equal in magnitude to the phase of the internal clock! In short, just as the frequency v is the rate at which full cycles of the internal clock pass by, so the energy hv is the rate at which the particle is "accumulating" action along its path.

Among other things, De Broglie's result $\lambda = h/p$ gave him a new way of explaining the Wilson-Sommerfeld quantization condition and the allowable Bohr orbits: from $\lambda = h/p$ and $\oint p\,dq = nh$ it follows that

$$\oint \frac{h\,dq}{\lambda} = nh \text{ or } \oint dq = n\lambda$$

In other words, the allowable periodic processes are those in which, on any cyclical path (such as Bohr's electron orbits), the particle traverses an exact *whole number of wavelengths* of the associated phase wave.

The rest of the history of wave mechanics (in which the principal actor was Erwin Schrödinger) need not concern us here. But in 1948 the American physicist Richard Feynman gave a new interpretation of quantum mechanics that emphasized the fundamental role played by action. In the Feynman interpretation, the principles of quantum mechanics take roughly the following form: anytime a physical process might occur (say, the motion of a particle from point A to point B), one examines *all* the possible ways (all the geometrically possible paths) by which that process can occur and calculates for each way or path the total action along it. One then divides that action by the Planck unit h to find the "phase" for that path (corresponding to the direction in which the sweeping hand of the particle's internal clock points as the particle reaches its destination). Comparing the phases of the various paths allows one to determine which paths interfere destructively (opposite phases) or constructively (same phases) with one another. By summing over all possible paths, one determines the probability that the process (motion from A to B) will occur at all.

But how in the world does this amount to a partial vindication of the surface teaching of Aristotle's *Physics* on the availability of a science of nature? It seems to me that what these scientists from Planck to Feynman have discovered is that nature *does* present itself to the inquiring mind of the physicist as dividing the world into fundamental units or elements (*stoicheia*). In a multitude of diverse phenomena—from black-body radiation, to the photo-electric effect, to atomic spectra, among others—nature just seems *intent* on revealing to us that *action* is the basic underlying stuff of which the events of fundamental physics consist, and that that stuff is composed of tiny *units* (though not in the sense of indivisible atoms) of action of equal magnitude. The magnitude of those units proves to be the Planck quantum of action h, and the number of such units of action that an object accumulates in traveling along a certain path from A to B gives us the phase associated with that path; and it is this phase that determines what contribution that path will make, in conjunction with all the other possible paths, to the probability that the object will in fact move from A to B.

Moreover, this method of "summing over all possible paths" has proved to be overwhelmingly successful in predicting the outcomes of the basic processes of nature.[9] It has been confirmed experimentally to an astonishing degree of precision (in the case of the fine structure constant, which governs the strength of interaction between elementary charged particles, the precision of the agreement between theory and experiment is currently one part in four billion), and there has been no evidence of a breakdown in this essential account of how nature works on that level. Thus, the Planck quantum of action proves to be deeply connected with the inherent periodicity of the de Broglie clock that measures the proper time of every bit of matter in the universe, and the discovered connection between action and time has laid the foundation for our present theoretical understanding of all the fundamental processes by which elementary particles interact with one another—the branch of physics known as *quantum field theory* or, alternatively, as the Basic Model. May we not conclude that the quantum of action is in a real sense the basic building block or element (*stoicheion*) out of which all these fundamental processes of nature are composed? And, in the de Broglie postulate that every body of mass m has an intrinsic periodicity or clock of frequency $\nu = mc^2/h$, and thus an associated phase tmc^2/h, equal to the action accumulated along the body's history when counted in units of h, can we escape the conclusion that quantum mechanics has revealed to us a fundamental principle at work in all natural processes and making them, in some measure, intelligible to us?

* * *

Many other such *archai* and *stoicheia* of nature have been revealed by modern physics; but on the strength of these two examples alone, it seems to me reasonable to conclude that modern science—whatever it was at its origin, whatever turns and twists it took, whatever narrow methodology it adopted, however entangled it may yet be with the project of making man the master and possessor of nature—has nonetheless made discoveries that not only *empower* us to predict and control the natural world, but also *enlighten* us with principles and elements that make the nature of all things, in some measure, intelligible. How a science that was invented chiefly to bear fruit also ended up shedding such a piercing light on the nature of things is perhaps a question worth contemplating.

Chapter 16

Quantum Mechanics and Political Philosophy

Bernhardt L. Trout

Quantum mechanics has been utterly successful as a physical description of the sub-astronomic. It has led to unprecedented technological accomplishments, such as microprocessors, modern pharmaceuticals, and advanced materials. It has led to compelling theoretical insights into nature, such as the picture of the atom and the understanding of its properties, both isolated and as a part of compounds. It has introduced new concepts, of which any educated person has at least a passing knowledge: the wavefunction, the uncertainty principle, and many-worlds, to name a few. All countries, whatever their regime, respect quantum mechanics, and all major religions feel the need to demonstrate the consistency of their tenets with quantum mechanics. It has captured the hearts of the brightest of youths and continually captivates the greatest of scientists. As such, it is more widespread than Christianity in its heyday. Despite these successes, the connection between quantum mechanics and political philosophy (i.e., the attempt to answer "What is?" for each of the political things) is far from obvious. This study elucidates some of these connections. In doing so, its focus is on the theoretical mastery of nature, which itself is directly connected to the practical mastery of nature.

In this seventeenth investigation, I have begun with the motivations of the developers of quantum mechanics and some of its key aspects that emerged. Second, I have presented arguments for the importance of quantum mechanics to political philosophy from the viewpoint of a certain quantum mechanician. Third, I discuss responses to those arguments from the viewpoint of a student of political philosophy. Fourth, I present responses from the viewpoint of a different quantum mechanician to the student of political philosophy. Finally, I revisit the strengths of quantum mechanics and the issue of its connection with political philosophy.

The Origins and Development of Quantum Mechanics

> [I wish] to explain as clearly as possible the real core of the theory. This can be done most easily by describing to you a new, completely elementary treatment through which one can evaluate—without knowing anything about a spectral formula or about any theory—the distribution of a given amount of energy over the different colours of the normal spectrum using one constant of nature only and after that the value of the temperature of this energy radiation using a second constant of nature.[1]

Quantum mechanics was developed in the first half of the twentieth century as a response to certain inadequacies in nineteenth-century physics. A poster child of these inadequacies is the so-called "ultra-violet catastrophe," which is the failure of the Raleigh-Jeans law, called the "classical theory." That failure is shown in figure 1 below, in which the classical theory falls miserably short of explaining the blackbody emission of radiation in the ultra-violet region of the spectrum. The classical theory works at high wavelengths, but starts to miss the mark at lower wavelengths and in the end diverges (goes to infinity) as the wavelength gets very small, a catastrophically absurd result.

At the turn of the century, Max Planck was able to develop a theory that fit the experimental data well and was consistent with the logical understanding that at low wavelengths the theory predicts no radiation intensity and does not diverge. However, this theory incorporated the fact that the blackbody could only take on specific discrete energy levels; that is, it was quantized. The significance of the quantization and acknowledgment thereof still took some years, but the seeds had been planted, and a completely new flora was to grow across the earth.

From these initial approaches to correct the classical theory with a transformed approach to understanding radiation, a new approach that aimed to describe all of nature was developed. Within a couple of decades, Paul Dirac could state that with quantum mechanics, "the underlying physical laws necessary for the mathematical theory of a large part of physics and the whole of chemistry are thus completely known."[2] It appeared that the principles of

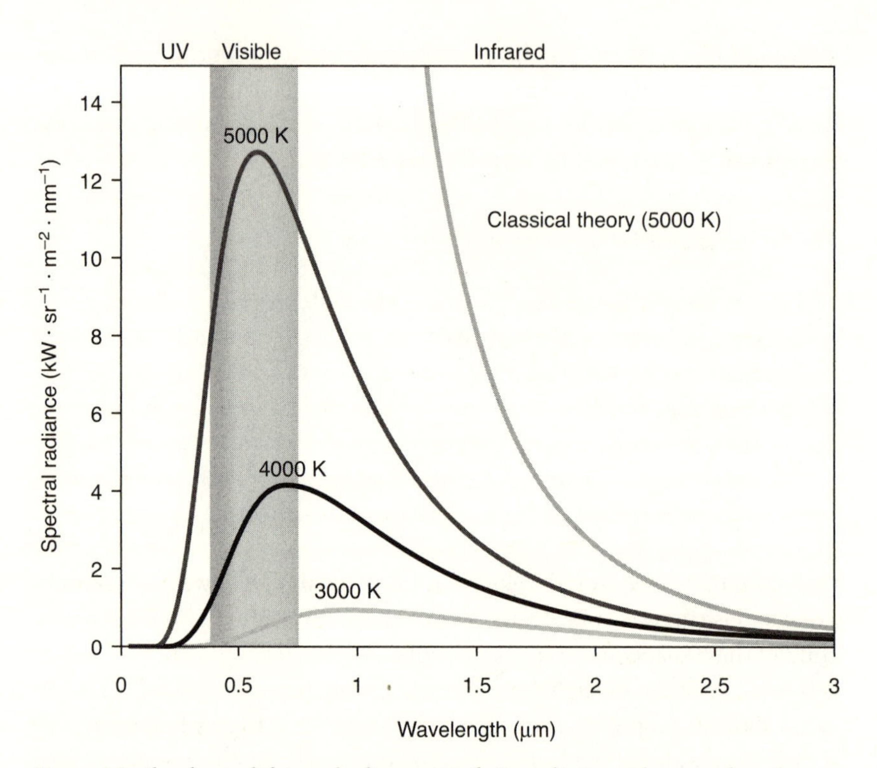

Figure 1. In the classical theory, bodies emit infinite radiation at low wavelengths, high frequencies (i.e., in the ultraviolet range). Quantum theory corrects this problem.

nature above the nuclear length-scale and below the astronomic length-scale (i.e., those that depend only on the electromagnetic force) had been found.[3]

In the subsequent decades, quantum mechanics was expanded to quantum field theory, a unified theory of three of the four fundamental forces: the weak nuclear, the strong nuclear, and the electromagnetic. Quantum field theory has been developed into the Standard Model, which is a comprehensive description of the "fundamental particles" of nature and considered to be a comprehensive description of the sub-astronomic.

During the same time in which quantum mechanics was developed, Einstein developed his theory of general relativity, which treats the astronomic, in which gravity, the fourth fundamental force, is important. The treatment of the very large with general relativity and the very small with quantum mechanics (or more broadly, quantum field theory) led to a search for a unified way to treat all length scales and forces, the Theory of Everything (TOE). Finding such a theory has, to date, eluded physicists. Nevertheless, physicists view quantum mechanics as part of the TOE, which is considered to be the comprehensive way of understanding nature, even though the TOE itself has not been articulated. Furthermore, the problems that quantum mechanics raises are fundamental in the sense that they deal with the building blocks that make up the manifest world. Those building blocks can be addressed with quantum mechanics without taking into account gravity, which is insignificantly weak in the microscopic world. Finally, despite billions of applications, there has never been found any instance in which quantum mechanics, when applied properly, fails. Quantum mechanics seems to be the true description of the sub-astronomic world.[4]

For those reasons, it is justified to treat quantum mechanics separately, while never forgetting that it is a part of a more comprehensive theory that aims to describe everything. There are many studies that address quantum mechanics for a lay audience and that cover to greater or lesser degree the gamut from the physical to the philosophical.[5] This paper does not aim to be comprehensive or to address adequately the physical features of quantum mechanics that make it utterly successful (predictive, universally respected, correct in all given applications) on the one hand, and utterly bizarre (e.g., irreducible randomness, non-locality,[6] and antiparticles) on the other. It does aim to consider the significance of the key features of quantum mechanics (within the presumed framework of TOE) for political philosophy.

A Quantum Mechanician's View of the Natural and the Political Things

> Since knowing imagistically and knowing scientifically occur for all investigations of which there are principles or causes or elements, from knowing these. . . .[7]

If the key to nature is knowing and understanding principles, causes, and elements, one should investigate what insight into these quantum mechanics might provide. The manifest world consists of things that are made of very small elements, "fundamental particles," that are indivisible.[8] These elements are distinguishable by specific properties that can be characterized. They interact with one another via inviolable principles, which are manifest as patterns to the senses. The principles themselves can be expressed in specific mathematical terms, the equations of quantum mechanics, which contain within them a small number of unchanging constants, "universal constants."

As these elements interact with each other, they combine and separate. When they combine, they form compounds which exhibit more complex patterns than the individual elements. They can combine to form large objects, macroscopic things, which are detectable by our senses. Our senses themselves are the results of interactions between us and other compounds, or between various parts that constitute us, all of which are governed by quantum mechanics.

Such a picture leads to a fully consistent explanation of the world in terms of mechanical processes. It provides, in principle, answers to all questions about nature, and it has recently provided answers to many basic questions that were hitherto unanswerable.[9] Why is there dew on grass in the morning? Because radiation dissipating from the grass cools it relative to the atmosphere and condenses the humidity in the air nearby. Why is the sky blue? Because the white light from the sun interacts with the atmosphere in ways that lead to blue light being separated out and reaching earth. Why are things hard or soft? Because electrons interact with each other and with atomic nuclei in ways that lead to different degrees of spatial response when macroscopic objects interact. And so forth. All of these can be quantified with a tremendous degree of accuracy via quantum mechanics.

Quantum mechanics presumes space and time. Within these, fundamental particles move and interact, combine and separate. Moreover, they can come into existence and go out of existence. All of these processes occur via the laws of quantum mechanics, such that patterns form when many of these processes occur, but a given process takes place within an irreducible randomness. There is no void per se, because everything is everywhere, albeit in a temporally transient, particular, non-uniform way following the principles of quantum mechanics. There are discrete things only because interactions between the things are weak compared with the interactions within them. For this reason, we are separate from the air around us, but as we breathe and oxygen in the air is incorporated into our body, it becomes part of us. It (and we) are technically everywhere at once, but because of all of the constant interactions around us, we practically occupy a given volume.

Things in the manifest world can be understood in terms of their parts via quantum mechanics. Those parts combine in ways that lead to things with completely new properties. For example, atoms—combinations of electrons, protons, and neutrons—have qualitatively different properties from their constituents. Molecules, which are combinations of atoms, have qualitatively different properties from their constituents. The same is the case for combinations of molecules (for example, H_2O molecules combined to form water) and also for rocks, hammers, and even for living things, plants and animals.

The motions (operations) of living things can be thought of as a consequence of "algorithms" in those things, analogous to computer algorithms. Those algorithms are a consequence of the organization of the fundamental particles that make up living things. Because of "decoherence"—that is, the interaction of particles that leads to their quantum behavior being averaged out, so that they exhibit classical behavior—these algorithms can lead to processes which do not appear to involve randomness. Nevertheless, there is always some degree of residual randomness in any process.

Living things, and in fact all things, interact with and react to their environment. In living things, those interactions can affect their algorithms via the senses. All of these physical encounters operate according to the principles of quantum mechanics, but the degree of complexity can be tremendous. For example, the physiological reaction to pain or pleasure involves so many interactions that enumerating them would take much, much longer than the lifetime of the universe. Particularly human things—beauty, justice, piety, etc.—are a result of the workings of these complex algorithms, which themselves are a product of interactions according to quantum mechanics.

Consciousness is also a manifestation of these algorithms, or something like them. It is currently far from being fully explained, but it ultimately results from quantum mechanical processes. As such, there is no fundamental difference between living and dead things (although clearly these things can be categorized differently). Just as the nineteenth-century notion that there is a separate, irreducible "living force" faded away into the history of the useless, there is no reason not to think that any remaining notions that there is a separate, irreducible "mind" will suffer the same fate, unless "mind" is just a shorthand for the algorithm formed from the interactions of particles that constitute the brain.[10]

Within the description of the world of quantum mechanics, something like all four Aristotelian causes are present. The material cause is a product of the properties of the compounds, which themselves are ultimately made of the fundamental particles, which themselves have specific properties. The efficient cause is the action of those complex algorithms that result from the combination of the fundamental particles. The formal cause is the pattern resulting from these combinations. The final cause is the specific properties resulting from combinations, especially in the complex algorithms that result, which direct things to certain ends. Plants grow, animals mature, and human beings strive for happiness. There is no natural place for non-living things, including the fundamental particles that move through the universe, but living things thrive in specific places conducive to them.

Regarding ethics and politics, they are important, but they are at best a necessary distraction from the essential work of studying the fundamentals of nature. Politics, in particular, is a necessary evil, but it is tolerable as long as it is willing to support quantum mechanics in praise and in funding. Philosophy, literature, and the arts are important, particularly to the extent that they help bring out the full implications of quantum mechanics.[11] Truth, however, is paramount, and the only way to obtain certain truth is via quantum mechanics.[12] Its success speaks for itself. Perhaps today, we need political philosophy to investigate final cause, but since in principle everything in the sub-astronomic world can be explained via quantum mechanics, political philosophy is at best secondary.

A Response to the Quantum Mechanician

> About those at the top . . . it's surely these who since their youth, first of all, don't know the way to the marketplace, or where's a court, councilhouse, or anything else that's a common assembly of the city.[13]

The political things are in the manifest world, and thus, political philosophy, to the extent that it focuses on investigation of the political things, tends away from the study of the microscopic and astronomic. The "algorithms" of human thought and action that lead us to study ethics and politics, in order to address such questions as how best to live and what constitutes the best society—even if resulting from indivisible, fundamental particles propagating and interacting via the principles of quantum mechanics—are qualitatively different from their material constituents, and so different that addressing these questions does not seem to necessitate understanding the microscopic processes and things that lead to these questions. Besides, although in principle those algorithms could be determined by solving the equations of quantum mechanics, a very rough and low estimate shows that it would require 10,000,000,000,000,000,000,000,000,000,000,000,000 years even to begin to determine them.[14]

Most importantly, the political things and, for that matter, all the things of the manifest world, our world, stand on their own, independent of their material constitution. A book, even if blank, is something above and beyond its matter. Beauty needs to be described on its own terms and cannot be reduced to the mathematical without the loss of an essential aspect of what it is. Even if we eventually discover the quantum mechanical formula for justice, quantum mechanics cannot tell us how we should best apply it. What function do we optimize? The greatest happiness, or the optimal punishment to reduce crime, or the maximization of dignity, to name a few possibilities? In other words, the political things are irreducible; they cannot be understood from understanding their constituents, however perfect that understanding may be.

Besides, quantum mechanics is the great-grandchild of the project of the mastery of nature, the Baconian-Cartesian project. It is therefore ultimately a product of political philosophy. That project itself is questionable, and instead of focusing on a derivative product of the project, one should focus on the project itself, with the origins and with the arguments for and against the project presented by the greatest thinkers. That quantum mechanics has led to tremendous technologies that affect almost every aspect of our lives only demonstrates the success of the Baconian-Cartesian project, and therefore that one should understand the project. The true understanding of the technologies and their success is understanding their origins.

Similarly, the theoretical underpinnings of quantum mechanics are tenuous at best. As a product of the modern scientific method, it is necessarily hypothetical. It cannot say, for example, that what it today considers to be the fundamental particles will never themselves be found to consist of parts. The fact that religions feel the necessity to demonstrate their consistency with quantum mechanics, and quantum mechanics does not deem it necessary to address religion, only demonstrates that religion is more comprehensive. After all, quantum mechanics cannot prove that miracles are false.

The success of quantum mechanics (i.e., an understanding of the world via application of the Baconian-Cartesian method) is only a way to reinforce dogmatically the original goal. It does not prove that the goal is the right goal or that the method can address all questions. Application of the method only leads to answers for questions which are suitable for the method to address.

That the method can be used to elucidate the atom does not mean that the method can elucidate everything. In particular, its success in one realm does not mean that the method can elucidate the political things, aside from perhaps transforming them to something else than they are and applying the method to those new things.

Because quantum mechanics was explicitly developed to address the microscopic, its application to the manifest world is tenuous at best. The development of quantum mechanics itself occurred in the macroscopic world. Moreover, if it truly has a significantly higher status than other theories, it cannot be a product of its own tenets: mechanism and randomness, homogenizing concepts which lead to no real hierarchies. On the other hand, quantum mechanics could be just a trick of deterministic processes that result in patterns indicative of quantum mechanics and its irreducible randomness, but that are not really random. As such, quantum mechanics becomes an illusion, a cosmic joke on its practitioners and supporters.[15]

There is at least one strong indication that quantum mechanics is not, in fact, the full picture of the microscopic world. It relies on mathematics, which is intrinsically unable to generate the randomness of quantum mechanics. It can only generate patterns that arise from a large number of presumably random processes.[16] If mathematics is part and parcel of quantum mechanics, quantum mechanics cannot even describe what it promises to describe. It is begging the question to use a method that is not suitable for randomness and then conclude that randomness is irreducible.

Above and beyond all of this, quantum mechanicians, no matter how successful they are, still need to address the questions of political philosophy. Quantum mechanics cannot even justify via quantum mechanics why one should study quantum mechanics. Such a justification would need to come from outside of quantum mechanics. This is sufficient to show that the quantum mechanician's scorn for the political things is rooted in insufficiently reflected-upon opinion. Thus, the right approach to understand the world is the approach of political philosophy, not the approach of quantum mechanics.

Applying the approach of political philosophy to the quantum mechanicians reveals the true understanding of what quantum mechanics is all about. For example, Max Planck reveals his aims: "I had always regarded the search for the absolute as the loftiest goal of all scientific activity"[17] and "We may say that the characteristic feature of the actual development of the system of theoretical physics is an ever extending emancipation from the anthropomorphic elements."[18] In order to meet this goal of the absolute, the chains of the anthropomorphic elements need to be removed. A human strives for human perfection by eliminating the human, an essential part of the whole, and as such develops a distorted conception of the whole.

Perhaps, however, the actual goal emanates from a deeper aim in the pursuit of modern physics. "Then we shall at once more correctly understand from that principle what we are seeking, both the source from which each thing can be made and the manner in which everything is done without the working of gods."[19] After all, ancient atomism and modern physics have in common that their approaches assume no divine action. Perhaps respect or prudence limits explicit statements of that aim, but modern physicists can make bold statements. Ludwig Boltzmann, for example, explicitly denigrates philosophy together with metaphysics on the one hand and belief in God on the other.[20] In doing so, Boltzmann shows that for him, modern physics becomes a substitute not only for philosophy and metaphysics, but also for transcendence and providence.

A Counter-Response

> He said that true opinion with speech was knowledge, but true opinion without speech was outside of knowledge, and of whatever there is not a speech, these things are not knowable.[21]

Gaining an understanding of the broad significance of quantum mechanics to thought is a challenge for many reasons, not the least of which is that it is unclear where to start. There are no key texts of quantum mechanics that can reveal a deep understanding of its appeal, aims, and promises. There are, however, speeches and writings of quantum mechanicians that can be studied to reveal insights into quantum mechanics. Understanding those speeches and writings requires working through them, including the complex mathematics that is an integral part of them. Learning this mathematics, however, takes time and effort, akin to learning a language. Nevertheless, this effort can be rewarded by understanding quantum mechanics' utterly appealing power for gaining understanding.

For example, shortly after Schrödinger's posing of his eponymous equation in 1926, which led to an understanding, for the first time, of the hitherto elusive atom (including the ability to calculate the quantized energy levels of the hydrogen atom, in addition to, in principle, those of any other atom or molecule), Dirac aimed to make Schrödinger's equation consistent with special relativity. In doing so, he posed his own eponymous equation:[22]

$$\left[\left(-\frac{W}{c}+\frac{e}{c}A_0\right)^2+\left(-p+\frac{e}{c}A\right)^2+m^2c^2\right]\psi=0$$

The consequence of doing this is that the wavefunction, ψ, must be a vector of specific 4×4 matrices. This means that ψ contains the two spins of the electron, ↑ and ↓, but each with both positive and negative energies. Negative energies seem impossible, however, like having a negative baseball. Various explanations ensued, but they were not fully satisfactory. Four years later, a satisfactory explanation was given when the antielectron, or positron, was found experimentally. Thus, the insight gained in solving a mathematical problem led to gaining an astounding new insight into the nature of the world. It is as if solving the equations of motion for a baseball led to the prediction of an antibaseball which was then found in the world, and which, when it interacts with a baseball, cancels it out in a flash of light.

Such success demonstrates that the principles of quantum mechanics are unmatched in their comprehensiveness, insight, and powers of prediction. As such, it is seductive, perhaps even more so being counter-intuitive and revealing truths that can, at best, be described as bizarre. Aside from the result that things constantly come into and go out of existence, other results of quantum mechanics are that each thing can be described as having associated with itself a wavefunction, which leads to all measurable properties but is itself unmeasurable; there is inherent, irreducible randomness that governs every process; actions occur non-locally; and macroscopic things are localized only because interactions lead to wavefunction collapse.

Quantum mechanicians, however, are not content with merely solving problems or gaining insight. They aim to understand the fundamental aspects of nature. (And in fact, they are in the mainstream of speculative philosophy today.) The subtitles in two of the references mentioned previously are *My Quest for the Ultimate Nature of Reality* and *Finding Nature's Deep Design*.[23] Having said that, quantum mechanicians are open to other approaches. Wilczek states bluntly, "Science isn't everything, thank goodness."[24] Thus, quantum mechanicians can hold the political things as important and acknowledge that they must be studied differently from the proper objects of the modern scientific method. They can acknowledge the limits of the scientific method, while still holding that nature reveals herself, not completely, but to a great extent and marvelously via the modern scientific method. They also acknowledge that healthy politics is necessary for modern science.

For those who seek the truth, it matters that atoms exist and what they are. It matters that the earth is not the center of the universe, that the species are not eternal, and that there are intrinsic limits to the mastery of nature (due to the laws of thermodynamics and quantum mechanics, for example). These also matter for political philosophy. Call them reigning dogmas if you wish, and underscore the fact that modern science is foundationally hypothetical. But at the least, one must say that these dogmas are healthier to politics than rulers being gods and priests interpreting omens for important decisions. Quantum mechanics is an immense achievement of the intellect and, by its success, demonstrates that nature reveals herself at least in part to those who truly wish to know her secrets.

Quantum Mechanics and Political Philosophy Within the Whole

> Philosophy, as quest for wisdom, is a quest for universal knowledge, for knowledge of the whole.[25]

The whole seems to consist of both political things and non-political things. However, the political and the non-political are not separable in a straightforward way. Quantum mechanics, which at first sight may seem non-political, has tremendous political consequences. On the other hand, quantum mechanics, as a key part of natural philosophy, might be able to demonstrate the primacy of natural philosophy over political philosophy. Beauty, justice, piety, and other things of political philosophy could in fact be merely a result of mechanical processes described by quantum mechanics.[26]

That the consequences of mechanism might be devastating to political philosophy does not mean that it should be dismissed. That the political things might be reducible to the non-political should be taken seriously. The possibility that the whole might, in the end, be ultimately homogeneous, a space-time continuum into and out of which things fluctuate, is a possibility that political philosophy must face with courage.

Even if quantum mechanics is a product of a certain stream of thought within political philosophy, the Baconian-Cartesian project, this does not mean that the products of quantum mechanics are fully reducible to political things. For there seem to be things within the whole that simply are, independent of the political things. Furthermore, in quantum mechanics, we have, for the first time, a non-reductionist but mechanistic approach that is at least open to wholes as not simply the products of the combining of their microscopic constituents. There is much left to be done to work out this possibility, but at least in principle, the fact that there are underlying particles in nature that constitute each thing is, via quantum mechanics, compatible with the notion that there are independent wholes that must be understood on their own terms.

If, however, the political is not reducible to the mechanistic, what is the relationship between quantum mechanics and political philosophy? Philosophy addresses the whole; it addresses what each thing is within and including the whole. If political philosophy is that part of philosophy that addresses political things, it must do so within the whole. The teachings of quantum mechanics have both political and nonpolitical implications. How those relate to each other is an important subject for philosophy, the quest for knowledge of the whole. This is the case whether quantum mechanics is a reigning dogma or the true way to investigate nature or something in between.

If the things that make up the whole are ultimately heterogeneous in kind, they need to be studied in different ways. The Baconian-Cartesian method seems to be apt for certain kinds of things, but not others. It is, however, unclear a priori where the boundaries of applicability lie. Atoms can be studied as political phenomena, while beauty can be studied scientifically. However, in both of these approaches, something is lost in the object studied. Understanding this is part and parcel with countering dogmatism undogmatically.

If the Socratic approach is to consider "what is" each of the beings, Socrates would not exempt quantum mechanics and its objects from his analysis. Quantum mechanics is a key part of the success of natural science on the one hand, and it is itself a major political (or trans-political) phenomenon on the other. Socrates would investigate what quantum mechanics is.

Notes

Chapter 1. Machiavelli and the Discovery of Fact

1. John Locke, *Second Treatise of Government*, paragraphs 25–27, 32, 43, in the context of ownership of property; and Locke's *An Essay Concerning Human Understanding*, book 1, against innate principles or ideas, both practical and speculative. Locke's anti-nature epistemology accompanies and accommodates his anti-nature politics; the first is skeptical, the second is assertive.

2. For example, at the very end of Aristotle's *Politics*.

3. Citations from Machiavelli's *Discourses on Livy* are to book, chapter, and paragraph, using the translation of Harvey C. Mansfield and Nathan Tarcov (Chicago: University of Chicago Press, 1996), and cited in the text as "*D*." Citations from *The Prince* are to chapter and page in the translation of Harvey C. Mansfield, 2nd ed. (Chicago: University of Chicago Press, 1998), and cited in the text as "*P*." Citations from the *Florentine Histories* in the translation of Laura Banfield and Harvey C. Mansfield (Princeton, NJ: Princeton University Press, 1988), and cited in the text as "*FH*."

4. Aristotle, *Nicomachean Ethics*, 1094a1–27, 1094b7–12; cf. *Politics* 1252a1–7.

5. Pierre Manent, *Naissances de la politique moderne* (Paris: Payot, 1977), 35.

6. Aristotle, *Nicomachean Ethics*, VI, 1144a23–29. In *The Prince*, astuteness is associated with greatness; in *P* 15.62 it is paired with honesty in the contrast of human "qualities," and Severus (the model criminal founder of *P* 19.79, 82) is said to be a "very astute fox" as well as a "very fierce" lion. Prudence seems to be more public, less hidden, than astuteness (*D* 1.6.4, 1.41); in *P* 23.95, a prince is said to be "very prudent" who by chance might "submit himself to one alone who might govern him in everything, who is a very prudent man." Machiavelli may consider himself very prudent not to allow himself to be considered, in his own case, very astute. But this does not amount to Aristotle's distinction between prudence and cunning. One might say that prudence is the teaching of astuteness (see the "astuteness of Hercules" in *D* 2.12.2). "Human astuteness and malignity" may or may not have a limit (*D* 2.5.2).

7. See David Wootton, *The Invention of Science* (New York: HarperCollins, 2015), 252–254; Harvey C. Mansfield, "Machiavelli's Enterprise," in *Machiavelli's Legacy*, ed. Timothy Fuller (Philadelphia: University of Pennsylvania Press, 2015), 26. There are three instances of imagining in *The Prince* and eight in the *Discourses*, all critical of imagination (including two on religious imagination, one pagan, one Christian) except for one. That one, finding imagination useful, connects it to "science," in the only chapter in either book using that term (D 3.39.1–2).

8. See especially references to "the whole world" in *P* 3.9, 19.81; *D* 3.11.2, cf. *D* 2.2.2 (the "honor of the world"). Also "knowing the world" (*P* 18.70; *D* 3.31.3) and not knowing it (*D* 1.38.3, 3.31.3). And "worldly things," *D* DL.3.

9. Forms are imposed on matter rather than recognized in it; see *P* 6.23, 26.102, 104. See Leo Strauss, *Thoughts on Machiavelli* (Glencoe, IL: The Free Press, 1958), 337n118.

10. The word "sect" (*setta*) occurs once in *The Prince* and twenty times in the *Discourses*, seven times in one chapter (*D* 2.5). On the prince as prophet, see *P* 6.28; and for the distinction between armed and unarmed prophets, we are told that a prince can become armed if he knows the "art of war" (*P* 14.58, together with *P* 13.56 on David and Goliath).

11. With this insult I excuse my scholarly friends: Erica Benner, *Machiavelli's Ethics* (Princeton, NJ: Princeton University Press, 2009), ch. 4; John P. McCormick, *Machiavellian Democracy* (Cambridge: Cambridge University Press, 2011), 4; Maurizio Viroli, *Machiavelli* (Oxford: Oxford University Press, 1998), 24.

12. Justice is identified with necessity in *P* 26.103; *D* 1.2.5, 3.1.2.

13. The unnamed orator is surely Michele di Lando, who is identified and praised to the skies for his goodness (*bontà*) by contrast to the "Duke of Athens" (in *FH* 3.16–18), goodness that "never allowed a thought to enter his mind that might be contrary to the universal good." One wonders whether goodness so ambitiously defined might yet be compatible with the unnamed orator's urging of wickedness.

14. "And it should be considered that nothing is more difficult to handle, more doubtful of success, nor more dangerous to manage, than to put oneself at the head of introducing new orders" (*P* 6.23; cf. *D* 3.35.1).

15. See Marsilius of Padua on the notion of "voluntary poverty," *Defender of the Peace* 2:12–14. The word "sect" occurs thirteen times in this work.

16. "Economism is Machiavellianism come of age"; Leo Strauss, *What is Political Philosophy*? (Glencoe, IL: The Free Press, 1959), 49.

17. Thus what is necessary can also be understood as the "more honorable part" (*D* 1.7.4).

Chapter 2. The Place of the Treatment of the Conquest of Nature in Francis Bacon's *On the Wisdom of the Ancients*

1. *Great Instauration*, preface, paragraph 4; *Novum Organum* (cited in the text as *NO*) I.86, 89; *Of the Proficience and Advancement of Learning, Divine and Human* (cited in the text as *AL*) I.iii.4; II.xvii.6–7. Bacon preferred to speak of "reason," "mode," and "way." I have made occasional use of Heidi Studer's unpublished and unfinished bilingual edition of *On the Wisdom of the Ancients*; or perhaps the title should be rendered "On the Wisdom of the Old." The word in question is a form of *vetus*, which is the standard word for "old"; in various places in the book, Bacon does use *antiquus*, too, which is the more obvious equivalent of "ancient." Studer's unpublished dissertation is the best effort known to me to articulate the argument and the action of the work as a whole; I have also found helpful the work of Howard B. White, Timothy Paterson, Robert K. Faulkner, Tom van Malssen, Jean-Pierre Cavaille, Rhodri Lewis, Sophie Weeks, and Robert M. Schuler.

2. The penultimate fable, "Metis, or Counsel" (§30), may be said to belong to that sequence, especially since it contains an "*arcanum* of empire," but this "monstrous and at first glance most tasteless fable" is in fact a transition to the final religious-philosophic parable-fable on the Sirens; it suggests that Bacon swallowed the "counsel" of Christianity to give birth to his project,

an "armed Pallas" (wisdom), or a fighting fist, instead of an open and irenic palm (see the first epigraph above).

3. For another way out of the experience of this chaos, see Stuart D. Warner's essay on Descartes (chap. 4 of present volume).

4. Not using the word "myth," Bacon draws a distinction between fables and parables and characterizes the first eighteen tales only as fables, never as parables. It is only in the crucial section on the conquest of nature that Bacon begins to speak of the stories as parables, as well as characterize sections of the stories as parables; and, to indicate the significance of the distinction, the final "chapter" of the work, "Sirens, or Pleasure," is *both* a fable and a parable, suggesting a synthesis of ancient fables and Christian parables (a resurrected and Baconianized Orpheus). "Prometheus" is initially characterized as a fable, but then gets called three times a parable. The only "pure" parables are "Daedalus, or the Mechanic" (§19), the first in the conquest of nature sequence; "Nemesis, or Vicissitude of Things" (§22), on the conquest of fortune, but more importantly on the harshness of the beliefs of the "vulgar" in divine punitiveness; and "Scylla and Icarus, or the Middle Way" (§27), on the moral preparation required to prevent the Promethean anxiety of the scientist. In the parts of *WA* prior to the section on the conquest of nature, the only parable is the "manifest" one on Pluto's invisibility helmet in "Perseus, or War" (§7). Bacon's invisible war is manifest. On parabolic poesy, see *De Augmentis* (II.13, VI.2). On Bacon's manner, consider Ben Jonson's remark that "No man ever spake more neatly, more presly, more weightily, or suffer'd lesse emptinesse . . . in what hee utter'd" (*Timber: Or, Discoveries made upon Men and Matter: as they have flow'd out of his daily Readings; or had their refluxe to his peculiar Notion of the Times* [London, 1641], 101) with Aubrey's "Dr. Harvey tolde me that it [Bacon's eye] was like the Eie of a viper" (*"Brief Lives," chiefly of Contemporaries, set down by John Aubrey, between the years 1669 & 1696* [Oxford: Clarendon, 1898], 1:72).

5. "Scylla and Icarus" is one of only two "double-barreled" fables—the other one being "Actaeon and Pentheus"—one of them serving as a preface to the fable on "philosophy," "Orpheus," and the other as a preface to the fable on "science," "Sphinx," as well as reflections on the "social" image of science and thus on the way to promote it ("Sphinx").

6. Writing jointly to Sir William Cavendish, Fulgenzio Micanzio and the historian Paolo Sarpi expressed their admiration for Bacon's writings. In particular, *WA* had "opened a way to know that which Socrates meant in the Examination of Sciences, when he said that Poets utter great and brave things but they understand them not themselves." *WA* opens just as much "a way to know" what Bacon intended by the conquest of nature for the improvement of the human situation. In particular, the book allows us to see how that conquest or mastery relates to Bacon's moral and political thoughts, as well as his cosmological schemes. An account of Bacon's own myth, *The New Atlantis*, would show that it is a poetic utopia, a psychagogic device, apparently perfect and happy but impossible; cf. Tobin L. Craig, "On the Significance of the Literary Character of Francis Bacon's *New Atlantis* for an Understanding of His Political Thought," *Review of Politics* 72, no. 2 (April 2010): 213–239. The "New Atlantis" itself has poetry about it, too (an ancient-sounding solemnity), but it is a much more practical project—technologically empowered, imperial, colonizing. The new technological society, the New Atlantis, can no longer remain closed. A fundamental rejection of the political solution of the ancient is implied. Bacon's thought thus has fundamental unity since the New Atlantis (as *distinguished* from Bensalem, but emerging through a study of the impossibilities on Bensalem) is the new dynamic political structure which corresponds to the new science. Bensalem, moreover, is an image that

can motivate and mobilize the young and the daring. Bensalem is a psychopomp-ship conveying us to the New Atlantis, which, however, is never described explicitly. This would solve the fundamental interpretive problem of the title of the work as well as of the perhaps subtitle: "unfinished work." What is unfinished and yet to come is the new global technological regime. But on the fundamental issues—Bacon's correction of the ancients on technology; the necessity of his new rhetoric; the fate of virtue, freedom, or philosophy in a technological society; the utility of having the New Atlantean Bensalem as a motivating, moving target; the nature of the marriage between Bensalem and the New Atlantis; and the new conception of the relation between science and society—on all of these questions, *De Sapientia Veterum* is the authoritative book.

7. In the correlative third tale from the end, "Proserpina"—in a part of the fable specifically designated as a parable—conservation and restoration of bodies is said not to be expected from "some medicine or some simple and natural mode."

8. Orpheus's fault is not that he tried scientifically to achieve immortality ("Proserpina, or Spirit," §29, "ancients seemed not to despair . . . to prolong life"; "Prometheus," §26, "the ancients did not despair about modes and medicines about the retardation of old age"). But his natural science was not Machiavellian enough and did not sufficiently go back to first, "common principles" (the fable on an "*arcanum* of nature," "Deucalion, or Restitution," §21).

9. Even in *WA*, in "Prometheus, or the State of Man" (§26): against a combination that leads to *religio haeretica* and *philosophia commentitia*. Philosophy and religion "commixed together . . . will make an heretical religion, and an imaginary and fabulous philosophy" (*AL* II.vi.1); or the attack on *philosophia phantastica* and *superstitiosa* in *NO* I.63 and 65; see also the Ixion fable in *AL* II.viii.3 on vaporous, imaginary "science" versus "laborious and sober inquiry of truth." Strikingly, Bacon's student Hobbes has a political interpretation of Ixion's "copulation" with a cloud, in which Ixion stands for private judgment (i.e., for Socrates [preface to *De Cive*]). Cf. Kennington's unpublished "Bacon's Concept of Mastery of Nature": "final causality is the providence of the ancient contemplative philosophers"; "desire for power moderates eros for knowledge"; "mastery of nature moderates philosophy."

10. We hear in a casual remark in "Pan" that nature is matter.

11. Making man "like the center of the world as regards final causes" and making a virtue of ingratitude to God ("Prometheus"). "Prometheus" is a combination of "human nature" and "the author and magistrate" of human nature; that is, a combination of the aimless mixing up of elements into the human "microcosm" and "self-made man."

12. See Herman Melville, *Pierre, or the Ambiguities*, ed. Harrison Hayford, Hershel Parker, and G. Thomas Tanselle (Evanston, IL:, Northwestern University Press and The Newberry Library, 1971), 7:135–136).

> Herein lies an unsummed world of grief. For in this plaintive fable we find embodied the Hamletism of the antique world; the Hamletism of three thousand years ago: "The flower of virtue cropped by a too rare mischance." And the English tragedy is but Egyptian Memnon, Montaignised and modernised; for being but a mortal man Shakespeare had his fathers too. Now as the Memnon Statue survives down to this present day, so does that nobly striving but ever-shipwrecked character in some royal youths (for both Memnon and Hamlet were the sons of kings), of which that statue is the melancholy type. But Memnon's sculptured woes did once melodiously resound; now all is mute. Fit emblem that of old, poetry was a consecration and an obsequy to all hapless modes of human life; but in a bantering, barren, and prosaic, heartless age, Aurora's music-

moan is lost among our drifting sands, which whelm alike the monument and the dirge. (VII.vi)

13. On the capacity to induce pity and belief, see the characterization of Perkin Warbeck in *History of Henry VII*, where he is described as

> a youth of fine favour and shape; but more than that, *he had such a crafty and bewitching fashion both to move pity and to induce belief*, as was like a kind of fascination and inchantment to those that saw him or heard him. Thirdly, he had from his childhood been such a wanderer, or (as the King called it) was a landloper, as it was extreme hard to hunt out his nest and parents; neither again could any man, by company or conversing with him, be able to say or detect well what he was; he did so flit from place to place. (emphasis added)

> He . . . was a finer counterfeit stone than Lambert Symnell; better done, and worn upon greater hands; being graced after with the wearing of a King of France and a King of Scotland, not of a Duchess of Burgundy only. And for Symnell, there was not much in him, more than that he was a handsome boy, and did not shame his robes. But this youth (of whom we are now to speak) was such a mercurial, as the like hath seldom been known; and could make his own part, if any time he chanced to be out.

14. See "Actaeon and Pentheus" (§10) on the dangers of incircumspection.

15. In a remarkable section in the second book of *AL*, Bacon denies in the same breath that man is the microcosm and asserts that man is in the image of God, leaving the world without the honor of being in the image of God. The adoption of this assertion would explain at the same time the "idealistic" view that "the subject is the substance" and the radical skepticism about the intelligibility of the world. It is permissible to remake the world since it is not dignified.

16. "Nemesis" (§22), the fable on the relation among fortune, providence, and righteous indignation, and "Achelous, or Battle" (§23), the "repetition" of war-fable §7 but applied to the war against nature, may appear to be digressive but in fact are not: §22 is about the architecture of fortune, and its limits, but also about the vulgar's envy of the powerful and successful and the consequent imaginary personification of a goddess that will deliver their comeuppance, while §23 is about the battle against the multiform river of fortune and the procurement of the horn of plenty.

17. This is why *NO* I.129 holds out the prospect of global, in a way transpolitical, fame for them.

18. Though Bacon omits the silver settings, the image may also refer to Maimonides's reading (in the Introduction to *Guide of the Perplexed*) of a verse from *Proverbs* 25:11 ("a word fitly spoken"): it is "a saying uttered with a view to two meanings is like an apple of gold overlaid with silver filigree work having very small holes"; Ralph Lerner, "On Speaking in the Language of the Sons of Man," *Proceedings of the Israel Academy of Sciences and Humanities* 8, no. 1 (2000): 5–18.

19. Cf. §5 and §16.

20. "Of Seditions and Troubles": "The politic and artificial nurturing and entertaining of hopes, and carrying men from hopes to hopes, is one of the best antidotes against the poison of discontentments. And it is a certain sign of wise government . . . when it can hold men's hearts by hopes, when it cannot by satisfaction." Once the God of science also reaches its vicissitudinous limit, further ploys will have to be devised to keep stringing the people along: "In civil actions, he is the greater and deeper politique that can make other men the instruments of his will

and ends, and yet never acquaint them with his purpose, so as they shall do it and yet know not what they do, than he that imparteth his meaning to those he employeth" (*AL* I.vii.7).

21. Descartes notebook, end, Leo Strauss Archive, Box 19, Folder 1 (1941); Strauss mentions Empedocles in this context as the one among the ancients to adopt that strategy.

22. On necessity (cosmological and political, and the ineffectiveness of oaths), see "Styx, or Treaties" (§5). In *AL* II.x.7, Epicurus becomes inebriated when he drinks the Stygian water. Far from conquering it, Epicurus embraces it as an escape from a disease. As Bacon explains, because Epicurus was diseased, he did not taste the bitterness of the Stygian waters. In the *De Augmentis* version of "Pan," there are two additions concerning Epicurus, both in connection with the vulgar. First, the statement that fortune is the daughter of foolish vulgar and favored with "lighter philosophers"—such as Epicurus who bends his natural philosophy to serve his moral interest in *euthymia*, refusing to entertain any questions that might disturb his conscience. In the expanded explanation of "panic terrors" (*WA* §6), however, Epicurus is first associated with the vulgar but is then favorably quoted as saying: "It is not profane to negate the gods of the vulgar but to apply to the gods the notions of the vulgar."

23. "Prometheus" is the only section of the book in which a cognate of Christ or Christianity appears.

24. Bacon's "stooping" is that of a hawk or a falcon in the double meaning used by Pope: "Say, will the falcon, stooping from above, / Smit with her varying plumage, spare the dove?" and "That not in fancy's maze he wondered long, / But stooped to truth and moralized his song" (Alexander Pope, *Essay on Man*, Epistle III, lines 53–54 and "Epistle to Dr. Arbuthnot," lines 340–341).

Chapter 3. Hobbes on Nature and Its Conquest

1. All parenthetical citations in the text are to Hobbes's works, referred to by the following abbreviations: *Critique du De Mundo de Thomas Hobbes*, ed. Jean Jacquot and H. W. Jones (Paris: Vrin, 1973), cited as "*Anti-White*"; *De Cive*, Latin Version, ed. Howard Warrender (Oxford: Clarendon Press, 1983), cited as "*De Cive*"; *De Corpore, Elementorum Philosophiae Sectio Prima*, ed. Karl Schuhmann (Paris: Vrin, 1999), cited as "*De Corp.*"; *De Homine*, in volume 2 of *OL* (see below), cited as "*De Hom.*"; *The Elements of Law Natural and Politic*, in *Human Nature and De Corpore Politico*, ed. J. C. A. Gaskin (Oxford: Oxford University Press, 1994), cited as "*Elem.*"; *The English Works of Thomas Hobbes*, ed. Sir William Molesworth (London: John Bohn, 1839–1845), cited as "*EW*"; the Latin *Leviathan*, ed. Noel Malcolm (Oxford: Clarendon Press, 2012), cited as "Latin *Lev.*"; *Leviathan*, ed. Noel Malcolm (Oxford: Clarendon Press, 2012), cited as "*Lev.*" by chapter and paragraph number; *Thomae Hobbes Malmesburiensis Opera Philosophica quae Latine Scripsit*, ed. William Molesworth (London: John Bohn, 1839–1845), cited as "*OL*."

Translations from Hobbes's Latin works are my own. In quotations from both Hobbes's Latin and English works, I have modernized capitalization, spelling, italics, and punctuation, except where doing so might obscure an important aspect of Hobbes's meaning or emphasis. Unless otherwise noted, all italics or other forms of emphasis are in the original texts.

2. See Leo Strauss, *The Political Philosophy of Hobbes: Its Basis and Its Genesis*, trans. Elsa M. Sinclair (Chicago: University of Chicago Press, 1963), 10–19, 132–135; Peter J. Ahrensdorf, "The Fear of Death and the Longing for Immortality: Hobbes and Thucydides on Human Nature and the Problem of Anarchy," *American Political Science Review* 94, no. 3 (2000): 581–582; and Vickie B. Sullivan, *Machiavelli, Hobbes, and the Formation of a Liberal Republicanism in England* (New York: Cambridge University Press, 2004), 94–95.

3. See Jean-Jacques Rousseau, *The Second Discourse*, in *The First and Second Discourses*, ed. Roger Masters, trans. Roger and Judith Masters (Boston: St. Martin's Press, 1964), 113–115.

4. Compare René Descartes, *Principles of Philosophy*, in *The Philosophical Writings of Descartes*, trans. John Cottingham, Robert Stoothoff, and Duguld Murdoch (New York: Cambridge University Press, 1985), 1.53, 1.63, 2.4–11, 2.64.

5. Leo Strauss, *Natural Right and History* (Chicago: University of Chicago Press, 1953), 174.

Chapter 4. Devising Nature

1. My thinking about Descartes owes a great deal to conversations with Ronna Burger, Hannes Kerber, Thomas Merrill, Martin Sitte, Charlotte Thomas, Bernhardt Trout, and especially Svetozar Minkov.

2. Of course, strictly speaking, it begins with the title and the title page. The *Discourse* was published anonymously, although it was common knowledge that Descartes was the author. Concerning Descartes's "anonymous" writings, especially the letters that appear at the start of *The Passions of the Soul*, cf. Hiram Caton, "Descartes' Anonymous Writings: A Recapitulation," *Southern Journal of Philosophy* 20 (1982): 299–311.

3. All translations unless otherwise noted are by the author. References to the *Discourse on the Method* other than from the "Synopsis" will be cited parenthetically by part and paragraph number. As stated above, the *Discourse* consists of sixty-five paragraphs. However, the comprehensive Adam-Tannery edition changed that division, dividing the ninth paragraph of Part V into three, presumably because of what the editors thought to be excessive length. Following its publication in 1902, that edition of the *Discourse* established itself as the French text of choice—it served as the textual base, for example, for Étienne Gilson's indispensable edition (1925)—and I have only been able to discover one significant French but no English language version published after it—Gilbert Gadoffre's edition, first published in 1941 by Manchester University Press—that adheres to the original and correct paragraph logic.

4. However, cf. David Lachterman, *The Ethics of Geometry: A Genealogy of Modernity* (New York: Routledge, 1989), 127–140. It should be noted, at least in passing, that the French *autobiographie* does not enter into the language until the nineteenth century.

5. Cf. *Discourse on Method*, II.4.

6. Cf. *Discourse on Method*, III.3.

7. Cf. the various essays in *The Book of Nature in Antiquity and the Middle Ages*, ed. Klaas van Berkel and Arjo Vanderjagt (Leuven: Peeters, 2005); and *The Book of Nature in Early Modern and Modern History*, ed. Klaas van Berkel and Arjo Vanderjagt (Leuven: Peeters, 2006); also cf. Elizabeth L. Eisenstein, *The Printing Press as an Agent of Change* (Cambridge: Cambridge University Press, 1979), 2:453–456.

8. Cf. *Discourse on the Method*, I.11.

9. Cf., for example, Ronald G. Asch, *The Thirty Years War: The Holy Roman Empire and Europe 1618–1648* (New York: St. Martin's Press, 1997). This war lasted for almost the entirety of Descartes's mature lifetime, and the final treaty ending it was signed only sixteen months before his death. On Descartes's specific interest in the war, cf. Timothy J. Reiss, "Descartes, the Palatinate, and the Thirty Years War: Political Theory and Political Practice," *Yale French Studies* 80 (1991): 108–145.

10. Cf. Giancarlo Maiorino, *Adam, "New Born and Perfect": The Renaissance Promise of Eternity* (Bloomington: Indiana University Press, 1987). Also, in thinking through these examples it should be noticed how Descartes inconspicuously couples them with some activity of mind—seeing,

imagining, and thinking. It should also be mentioned that the qualities of perfection in each example differ.

11. The examples about law—the prudent legislator, God, and the founder of Sparta—are particularly worth seeing through this lens, especially since an argument can be made that the first of these refers to Moses and, by implication, the Old Testament, and the second of these (in the midst of the Thirty Years War) to the New Testament.

12. Descartes's use of the language of "reformation" (II.2) in the paragraph immediately after he points to God having made the "ordinances" of the "true religion" (II.1) is not coincidental.

13. However, cf. the opening line of *Discourse on the Method*, III.6.

14. Cf. Martha Ornstein, *The Rôle of Scientific Societies in the Seventeenth Century* (New York: Barnard College, 1913), and Edgar Zilsel, "The Genesis of the Concept of Scientific Progress and Cooperation," *The Social Origins of Modern Science*, ed. Diederick Raven, Wolfgang Krohn, and Robert S. Cohen (Dordrecht: Kluwer Academic Publishers, 2003), 128–168.

15. Cf. Plato, *Phaedo* 99d.

16. Similarly, the "Synopsis" informs us that we will encounter "rules" of the method in Part II, but instead we find "precepts."

17. Connected to this, in the opening paragraph of Part I of the *Discourse*, when Descartes remarks *seemingly* apropos of nothing that the "greatest souls are capable of the greatest vices as well as the greatest virtues," he is also pointing to Galileo who, from the perspective of Christendom, has committed the greatest vices, but who, from the perspective of science and philosophy, has exhibited the greatest virtues.

18. Careful attention should be paid to Descartes's use of pronouns in the *Discourse* generally, but especially in his treatment of his provisional morality.

19. Cf. the last line of the *Discourse on the Method*, III.3.

20. Joseph Cropsey, "On Descartes' *Discourse on Method*," *Interpretation: A Journal of Political Philosophy* 1 (1970), 136.

21. Cf. Descartes's *The Passions of the Soul*, arts. 144–146.

22. Letter to Mersenne, March 1636, in *Œuvres philosophiques de Descartes*, ed. Ferdinand Alquié (Paris: Éditions Garnier Frères, 1963–1973), 1:516.

23. Cf. Gerhard Krüger, "The Origin of Philosophical Self-Consciousness," trans. Fabrice Paradis Béland, *The New Yearbook for Phenomenology and Phenomenological Philosophy* 7 (2007), 245.

24. Thus, instead of asserting, for instance, *tout ce que nous ne réussissons pas*, Descartes states *manque de nous réussir* (III.4), thereby failing clearly to indicate where the lack or failure is to be located.

25. Cf. Alfarabi, *The Attainment of Happiness* (§59–61), on the true philosopher; also, Leo Strauss, *Persecution and the Art of Writing* (Glencoe, IL: The Free Press, 1952), 17.

26. Cf. *Discourse on the Method*, III.6.

27. With respect to the end of Part III, the United Provinces (the Netherlands) fought intermittently with Spain for a period of eighty years, from 1568 to 1648, the last three decades of which are considered to be part of the Thirty Years War.

28. Cf. Leo Strauss, *Thoughts on Machiavelli* (Glencoe, IL: The Free Press, 1958), 297–299 (where one finds the suggestion that philosophy is a fighting business). Also, cf. Hans Speier's discussion of Grimmelshausen in his "Introduction" to the part on "Folly" in *Force and Folly: Essays on Foreign Affairs and the History of Ideas* (Cambridge, MA: MIT Press, 1969), 189–192.

29. In a letter to Balzac, May 5, 1631, Descartes describes his situation in the United Provinces as follows: "What other place in the rest of the world could one choose where all the conveniences of life, and all the curiosities that may be desired, could be found so easily as this one? What other country, where one can enjoy liberty so complete, where one can sleep with less inquietude, where there are always armies at the ready to guard us, where poisonings, betrayals, slanders are less known, and where there remains more of a remnant of the innocence of our ancestors?" *Œuvres philosophiques de Descartes*, 1:292.

30. Cf. Georges Van Den Abbeele, *Travel as Metaphor: From Montaigne to Rousseau* (Minneapolis: University of Minnesota Press, 1992), 61; and Jacques Lezra, *Unspeakable Subjects: The Genealogy of the Event in Early Modern Europe* (Stanford, CA: Stanford University Press, 1997), 124–125.

31. Cf. Descartes, *Rules for the Direction of the Mind*, 8.9. Kepler, Galileo, and Huygens, along with a whole tradition of French literature going back to at least *Roman de la rose*, were fond of anagrams as well. Also, cf. the first epigraph of this essay, which contains an anagram.

32. *Œuvres philosophiques de Descartes*, 1:495.

33. Has any principle of philosophy, much less a first principle, ever been so idiosyncratic?

34. This is only the second occurrence of "soul" in the *Discourse*, the first occurring, as we have already noted, in the opening paragraph of Part I: "The greatest souls are capable of the greatest vices as well as the greatest virtues."

35. *The Vulgate Bible: The Pentateuch* (Latin and English), Douay-Rheims translation (revised), edited by Swift Edgar (Cambridge, MA: Harvard University Press, 2010). For one reading of the Biblical meaning of the remark, cf. U. Cassuto, *A Commentary on the Book of Exodus*, trans. Israel Abrahams (Jerusalem: Magnes Press, 1967), 37–38.

36. On the parallels between The Book of Genesis and Part V of the *Discourse*, cf. Michael Davis, *Ancient Tragedy and the Origins of Modern Science* (Carbondale: Southern Illinois University Press), 76–77.

37. The two parts of *The World* were published posthumously, in 1662 and 1664, respectively.

38. Cf. *The World*, chap. 5. Also, see Jan Baptist Weenix's portrait of Descartes (Central Museum, Utrecht), in which Descartes is seated holding an open book, the verso page of which reads, *Mundus est fabula*—the world is a fable. For a discussion of this painting in the context of the *Discourse*, cf. Jean-Luc Nancy, "Mundus est Fabula," trans. Daniel Brewer, *MLN* 93 (1978): 635–653; and Jean-Pierre Cavaillé, *Descartes. La fable du monde* (Paris: Librairie Philosophique J. Vrin, 1991), 13–17.

39. Cf. *The World*, chap. 6.

40. Cf., for example, Jane E. Ruby, "The Origins of Scientific 'Law'," *Journal of the History of Ideas* 47 (1986): 341–359.

41. On Descartes's place in the history of mathematics in this respect, cf. E.J. Dijksterhuis, *The Mechanization of the World Picture*, trans. C. Dikshoorn (Oxford: Clarendon Press, 1961), 403–409.

42. On the large and interesting number of parallel statements in Bacon and Descartes, cf. André Lalande, "Sur quelques textes de Bacon et de Descartes," *Revue de métaphysique et de morale* 19 (1911): 296–311. Also, cf. Laurence Lampert, *Nietzsche and Modern Times: A Study of Bacon, Descartes, and Nietzsche* (New Haven, CT: Yale University Press, 1993), 145–160. On Bacon's conception of the good, cf. Svetozar Y. Minkov, *Francis Bacon's Inquiry Touching Human Nature: Virtue, Philosophy, and the Relief of Man's Estate* (Lanham, MD: Lexington Books, 2010), 23–42.

43. Cf. *Discourse on the Method*, I.3; I.14; III.5; V.10 (12 in most editions); VI.2.

44. Cf. *Discourse on the Method*, I.3.

Chapter 5. Montesquieu, Commerce, and Science

1. Montesquieu, *The Spirit of the Laws*, trans. and ed. Anne M. Cohler, Basia Carolyn Miller, and Harold Samuel Stone (Cambridge: Cambridge University Press, 1989), 338 (20.1). I have occasionally made slight changes in the translation.

2. Montesquieu, *Oeuvres completes*, ed. Roger Caillois (Paris: Gallimard Bibliothèque de la Pléiade, 1949), 1:54. For an English translation and commentary, see Diana Schaub, "Montesquieu and the Motives for Science," *The New Atlantis* (Spring 2008): 32–46.

3. In the earlier case we are told who is being cured ("peoples") as well as what they are being cured of.

4. Francis Bacon, *The Works of Francis Bacon, Volume 4, containing Novum Organum Scientiarum, Volume 1* (London: M. Jones, 1815), xii.

5. Joseph Cropsey, "On Descartes' *Discourse on Method*," in *Political Philosophy and the Issues of Politics* (Chicago: University of Chicago Press, 1977): 281, referring to Descartes's final paragraph of part 2.

6. In the "Author's Foreword," Montesquieu had put readers on notice to watch for neologisms and reappropriations: "I have had new ideas; new words have had to be found or new meanings given to old ones" (xli).

7. Complete list: Preface (five times), 3.6, 4.6, 10.4, 12.28, 12.29, 14.3, 15.3, 18.18, 19.27, 20.1, 23.28, 24.22, 25.13, 28.23, and 29.19.

8. Montesquieu, *Spirit*, xliii (Preface).

9. Even with the additional material, this is Montesquieu's final reference to "prejudice."

10. Montesquieu, *Spirit*, 618 (29.19).

11. See the opening lines of 29.1 "On the Spirit of the Legislator": "I say it, and it seems to me that I have written this work only to prove it: the spirit of moderation should be that of the legislator" (602).

12. Montesquieu, *Spirit*, 595 (28.41).

13. Montesquieu believes detail-minded, reflective readers will be able to "feel the certainty of the principles" only as they come to perceive the connections between and among "the truths" (xliv). There is something odd about Montesquieu's reassuring words, since at the same time that he encourages readers to reflect on the details, he indicates that he has not supplied all the details (since that would be "deadly boring"). In sum, then, he tells us that the path to certainty lies through reflection on deliberately suppressed details. No wonder there is so little agreement among the readers of Montesquieu about his ultimate views.

14. Instances of "*la nature des choses*": Preface, 1.1, 3.3, 6.6, 10.13, 11.19, 15.6, 15.10, 19.27, 22.12, 26.20, 26.24, 29.16, and 29.17.

15. Bacon, *New Atlantis and the Great Instauration*, ed. Jerry Weinberger (Wheeling, IL: Harlan Davidson, 1980), 7. See also the opening sentence of the "Procemium," where Bacon commits himself to test "whether that commerce between the mind of man and the nature of things, which is more precious than anything on earth, . . . might by any means be restored to its perfect and original condition" (1). See also p. 25 for the "expurgation of the intellect" necessary for the marriage of "the Mind and the Universe" to proceed.

16. Bacon, *Instauration*, 31.

17. Montesquieu, *Spirit*, 316 (19.15).

18. Ibid., xliv (Preface).

19. Both phrases allude to book 6 of Virgil's *Aeneid*.

20. Despite his struggles, Montesquieu closes the Preface with a telling self-assessment: "I do not believe," says he, "that I have totally lacked genius" (xlv).

21. Montesquieu, *Spirit*, xliv (Preface).

22. Ibid.

23. For a revealing example, see Montesquieu's explanation of the jurisprudence surrounding the "monstrous usage of judicial combat"—proof that "men, at bottom reasonable, place under rules their very prejudices" (563 [28.23]).

24. Montesquieu, *Spirit*, xliv (Preface).

25. The first two follow this pattern: "If I could make it so that _______, I would consider myself the happiest of mortals." The last one switches up the pattern: "I would consider myself the happiest of mortals if I could make it so that ______." The sequence is best read in reverse order, understanding the first two in light of the third.

26. Montesquieu, *Spirit*, xliv (Preface).

27. Ibid.

28. Montesquieu does start his exploration of the phenomenon of law from the widest possible cosmic vantage point. What is most remarkable about 1.1 is the speed with which he dispenses with these speculations (involving the law-bound status of the divinity). Perhaps human affairs do not depend on settling such matters.

29. Montesquieu, *Spirit*, xlv (Preface).

30. Ibid.

31. Ibid., xliii (Preface).

32. Ibid., xliv (Preface).

33. Ibid., xliii (Preface).

34. Ibid., xliv (Preface).

35. The difference is real, but less than appears on the surface.

36. Montesquieu, *Spirit*, 114 (8.3).

37. These "*compositions*" should be kept in mind while reading the preceding book 29, "On the Way to Compose the Laws."

38. Montesquieu, *Spirit*, 647–648 (30.19).

39. Ibid., 166 (11.6).

40. He never mentioned the name of John Locke, except in one snarky entry in his *Pensées* (#1105), about Locke's vanity, and another attributing a bon mot to him (#1205). For the English translation of the *Pensées*, consult Montesquieu, *My Thoughts*, trans. Henry C. Clark (Indianapolis, IN: Liberty Fund, 2012).

41. The details would take us on another treasure hunt perhaps better left to another time.

42. Montesquieu alludes to Descartes in *Persian Letters* #97 and in 1.1 of *The Spirit of the Laws*. There are also a number of direct mentions in his *Pensées* (#50, #775, and #1445). The following is typical: "Descartes taught those who came after him to discover his very errors. I compare him to Timoleon, who used to say, 'I am delighted that by means of me, you have obtained the freedom to oppose my desires'" (#775).

43. Montesquieu, "Motives for Science," 33.

44. Ibid.

45. Ibid., 34.

46. *Pensées* #1265.

47. Montesquieu, *Spirit*, 294 (18.18). *Pensées* #1265 has a stronger version of this statement: "Thus, there is nothing so dangerous as to hit the minds of the people too hard with miracles and prodigies. Nothing is more capable of generating destructive prejudices than superstition."

48. In the Preface, even before defining man as "that flexible being," Montesquieu had stressed mortality as a constitutive element of our being when he listed the three achievements that would make him "the happiest of mortals" (xliv).

49. Thomas Jefferson, "To Roger C. Weightman," in *The Portable Thomas Jefferson*, ed. Merrill D. Peterson (New York: Penguin Books, 1975), 585.

50. *Pensées* #1266. See also #112 on Spinoza and "the loss of the sublime."

51. There is actually one other use of "science." In a late chapter on paying court costs, Montesquieu speaks of "the science of evading the most just claims" (586 [28.35]).

52. Montesquieu, *Spirit*, 389 (21.20).

53. Ibid., 483 (25.4). Cohler gives this as "in theory."

54. Ibid., 39–41 (4.8).

55. However, the chapter is dense with references to ancient authors (Polybius, Plato, Aristotle, Theophrastus, Plutarch, Strabo, and Xenophon). Not all the ancients were athletes and fighters.

56. Montesquieu indicates that the ancients used other gentling agents as well, like erotic love. Blushingly, he lets us know that some ancient cities established homosexual connections with the aim of citizen bonding.

57. Montesquieu, *Spirit*, 41 (4.8).

58. This is why the title of the chapter on the musical warriors is "Explanation of a Paradox of the Ancients in Relation to Mores." Commercial mores are coherent rather than oxymoronic.

59. Montesquieu, *Spirit*, 338 (20.1 and 20.2).

60. Borrowing a famous phrase from a statesman of one of the "great peoples" formed as colonial offshoots of England, see 329 (19.27).

61. Montesquieu, *Spirit*, 328 (19.27).

62. In 20.2, Montesquieu contrasts the spirit of commerce with the spirit of hospitality "notable among bandit peoples" (339). Then in 23.29, he describes the monks as "a nation in itself lazy and one that maintained the laziness of others, because, as they practiced hospitality, an infinity of idle people, gentlemen and bourgeois spent their lives running from monastery to monastery" (456).

63. Montesquieu, *Spirit*, 456 (23.29). See also 7.6 for the Chinese emperor who had "an infinite number of the bonze monasteries destroyed" (102).

64. Ibid., 333 (19.27). The antecedent for the pronoun *chacun* is the tribe of historians, but the description seems meant to apply more generally to all citizens.

65. Montesquieu's other references to the reign of Henry VIII are negative: 207 (12.22) and 496 (26.3) on laws that violate the natural right to self-defense.

66. Montesquieu, *Spirit*, 316 (19.14).

67. Ibid., xliv (Preface).

68. Ibid., 316 (19.14). In Montesquieu's playful retelling of the story of Eve and her follower in taste (Adam), clothing takes the place of the apple.

69. Ibid., 312 (19.8). Montesquieu's contribution to the *Encyclopédie* was the entry on "Taste."

70. Ibid., 311 (19.6 and 19.8).

71. Ibid., 338 (20.1). Later, in 21.5, he says that "the history of commerce is that of communication among peoples" (357).

72. Ibid.

73. They are our version of the printing press and the compass (see *Pensées* #653).

74. Montesquieu, *Spirit*, 312 (19.8).

75. Ibid., 310 (19.5).

76. Bacon, *New Atlantis*, 71–82.

77. "Incentivize" is a neologism that Montesquieu might have accepted.

78. Montesquieu, *Spirit*, 313 (19.10).

Chapter 6. Bacon and Franklin on Religion and Mastery of Nature

1. Benjamin Franklin, *Writings*, ed. J. A. Leo Lemay, Library of America (New York: Literary Classics of the United States, 1987) (henceforth *Writings*), 181–184.

2. Francis Bacon, *The Works of Francis Bacon*, ed. James Spedding, Robert Leslie Ellis, and Douglass Denon Heath (Cambridge: Cambridge University Press, 2011) (henceforth *Works*), 6:428, 387.

3. *Writings*, 1359.

4. *Works*, 3:167–168.

5. Ibid., 6:452.

6. Ibid., 720–722.

7. *Writings*, 1017–1018.

8. See *New Atlantis and The Great Instauration*, 2nd ed., ed. Jerry Weinberger (Boston: Wiley Blackwell, 2017) (henceforth *New Atlantis*).

9. *Valerius Terminus* (Cap. 25), *Works*, 3:251; *New Atlantis*, 84–85.

10. But see "Of Unity in Religion," *Works*, 6:381–384.

11. *New Atlantis*, 73–76.

12. Ibid., 108, 109–110.

13. Ibid., 74; see 31–32.

14. Shaftesbury, *Characteristics of Men, Manners, Opinions, Times*, ed. Lawrence E. Klein (Cambridge: Cambridge University Press, 1999), 368, referring to *De augmentis scientiarum* (2.13), *Works*, 4:325.

15. This is Stanley Fish's term for Bacon's *Essays*; when read carefully, they consume themselves and undermine the reader's first, conventional expectations. *Self-Consuming Artifacts: The Experience of Seventeenth-Century Literature* (Berkeley: University of California Press, 1973).

16. "Of Death," *Works*, 6:379–380.

17. *Works*, 6:384–385.

18. For Franklin's take on this argument, see my *Benjamin Franklin Unmasked: On the Unity of His Moral, Religious, and Political Thought* (Lawrence: University Press of Kansas, 2005), 180–182.

19. "A Defense of Mr. Hemphill's Observations," *The Papers of Benjamin Franklin* (New Haven, CT: Yale University Press, 1960), 2:114.

20. *Writings*, 1358–1371.

21. Ibid., 145–151.

22. Philoclerus's argument is strikingly close in tone and argument to Hobbes's comments about spirits in parts three and four of *Leviathan*. First, Franklin uses the same example as Hobbes to explain how one can see a light (or hear a sound) that no one else does: the striking of one's own closed eye. See chap. 45 of *Leviathan* (ed. Richard Tuck [Cambridge: Cambridge University Press, 1996]). Moreover, Philoclerus says that we have little understanding of spirits and

how they might work, but never says they are incorporeal. Franklin surely knew that Hobbes argued that spirits could be very subtle bodies and admits that scriptural texts have "extorted from [his] feeble reason, an acknowledgement and belief, that there be also Angels substantial, and permanent." The issue for Hobbes was not the possible existence of spirits, but the impossibility of their being incorporeal. *Leviathan*, 278; see also 274, 310, 440, 442, 463.

23. *The Writings of Benjamin Franklin*, ed. Albert Henry Smyth (New York: Macmillan, 1907), 1:166.

24. *Writings*, 211–212.

25. Ibid., 1398, 242–248.

26. Ibid., 511–518.

27. Ibid., 242–244.

28. Ibid., 1157–1160.

29. "Conversation on Slavery," *Writings*, 646–653; "Letter to John Bartram," *Writings*, 843–844.

30. *Writings*, 1096–1098.

31. See my "On Bacon's New Atlantis," in *New Atlantis*, 148–151.

32. *Writings*, 956–960.

33. Ibid., 1421–1422. In the *Autobiography*, Franklin opines that the Indians are doomed and that the White man's rum will be the real weapon of choice.

34. "Remarks Concerning the Savages of North-America," *Writings*, 969–974.

Chapter 7. On the Supremacy of Contemplation in Aristotle and Plato

1. Bacon, *Of the Proficience and Advancement of Learning, Divine and Human*, I.v.11.

2. Consider *Politics* 1341a23 and especially 1341b32–1342a17. Leo Strauss has observed: "Now, fear and pity are precisely the passions which are necessarily connected with the feeling of guilt. When I become guilty, when I become aware of my being guilty, I have at once the feeling of pity toward him whom I have hurt or ruined and the feeling of fear of him who avenges my crime. Humanly speaking, the unity of fear and pity combined with the phenomenon of guilt might seem to be the root of religion." "Progress or Return?," in *The Rebirth of Classical Political Rationalism: An Introduction to the Thought of Leo Strauss*, ed. Thomas L. Pangle (Chicago: University of Chicago Press, 1989), 250.

Chapter 8. Xenophon and the Conquest of Nature

1. Compare this to the "Introduction" of *On Tyranny*, where Strauss maintained that "in contradistinction to classical tyranny, present-day tyranny has at its disposal 'technology' as well as 'ideologies'; more generally expressed, it presupposes the existence of 'science,' i.e., of a particular interpretation, or kind, of science. Conversely, classical tyranny, unlike modern tyranny, was confronted, actually or potentially, by a science that was not meant to be applied to the 'conquest of nature' or to be popularized or diffused" (Leo Strauss, *On Tyranny* [New York: The Free Press, 1991], 23). See also p. 27 where the conquest of nature means "in particular of human nature."

2. Leo Strauss, *What is Political Philosophy?* (Chicago: University of Chicago, 1988), 96.

3. The fragment as we have it from Empedocles's *On Nature* reads: "You shall learn all the drugs that exist as a defense against illness and old age; for you alone will accomplish all this. You shall stop the force of the untiring winds that rush upon the earth with their blasts and lay waste the cultivated fields. And again, if you wish, you shall lead the breezes back again. You shall

make seasonable dryness for men after the dark rains and change the summer drought for streams that feed the trees as they pour down from the sky. You shall bring back from Hades a dead man restored to strength" (Empedocles, frag. 111, in Kathleen Freeman, *Ancilia to the Pre-Socratic Philosophers* [Oxford: Oxford University Press, 1948], 111).

4. Strauss, *On Tyranny*, 25; cf. 56. See also Strauss, *Thoughts on Machiavelli* (Chicago: University of Chicago Press, 1984), 59, 139, 291–293.

5. Francis Bacon, *The Advancement of Learning* (Oxford: Benediction Classics, 2008), 15, 30, 60. Benjamin Franklin, *Autobiography* (New York: Longmans, 1916), 15; Benjamin Franklin, "A Man of Sense," *Pennsylvania Gazette*, February 11, 1734; Benjamin Franklin, "Self Denial Not the Essence of Virtue," *Pennsylvania Gazette*, February 18, 1734. See Jerry Weinberger, *Benjamin Franklin Unmasked* (Lawrence: University Press of Kansas, 2005), 18, 57, 136, 201.

6. *Cyro.* 1.6; *Mem.* 1.1.6–7. See Algernon Sydney in *Discourses Concerning Government*: "That is the best government, which best provides for war," Sec. 23.

7. *Mem.* 1.1.15. Louis-André Dorion, the editor of the outstanding revised edition of the Greek text, thinks these questions are merely rhetorical, since it is "easy to guess [Xenophon's] answers"; see *Xénophon, Les Mémorables*, ed. Louis-André Dorion (Paris: Les Belles Lettres, 2010), 1:62 n.43. But Xenophon uses the same technique not infrequently, and even in the very next chapter, where his answer is far from clear (*Mem.* 1.2.8, with 1.2.40–46, 2.1.26).

8. See James Tatum, *Xenophon's Imperial Fiction* (Princeton, NJ: University of Princeton, 1989), 147–157.

9. Herodotus, 1.174.

10. Ibid., 1.191.

11. See Empedocles, fragment 112 with S. Panagiotou, "Empedocles on His Own Divinity," *Mnemosyne* 36 (1984), 276–85.

12. Plato, *Apology* 23d, 28a, 38c.

13. Cicero, *Tusculan Disputations*, V.10–11.

14. Compare "*pollōn kai ōphelimōn*" at 4.7.5 with "*allōn pollōn te kai ōphelimōn*" at 4.7.3 for the difference between geometry and astronomy. Accordingly, Socrates does not claim that gods put any restrictions on the study of geometry (4.7.6).

15. *Cyro.* 1.1.3. Cf. Shakespeare, *Coriolanus*, 1.1.240.

16. "Fathers contrive moderation in their sons, teachers good learning in their pupils, and the laws justice in citizens by making them weep" (*Cyro.* 2.2.14; cf. 3.3.50).

17. See Plato, *Laws* 889d–890a, 967a–e.

18. Cf. *Genesis* 3:19.

Chapter 9. Lucretius on Rebelling Against the "Laws" of Nature

1. E.g., 1.148, 6.41. All references to the Latin text are to Lucretius, *De Rerum Natura*, trans. W. H. D. Rouse, rev. Martin Ferguson Smith (Cambridge, MA: Harvard Loeb, 2006). I have consulted this and other translations, but all translations are mine.

2. Jacob Klein, "On the Nature of Nature," in *Lectures and Essays*, ed. R. Williamson and E. Zuckerman (Annapolis: St. John's College Press, 1985), 232. Klein distinguishes between writers of the seventeenth century and "French writers of the eighteenth century": for the latter, laws do not relate to any legislation.

3. Cf. Leo Strauss, "Notes on Lucretius," in *Liberalism Ancient and Modern* (New York: Basic Books, 1968), 89, 91.

4. 1.849–858. Lucretius explicitly restates, in 1.915–920, that the atom's nature must be different.

5. By law or covenant, according to Cyril Bailey, *Titi Lucreti Cari De Rerum Natura Libri Sex*, 3 vols. (Oxford: Clarendon, 1998), ad loc. 1.586, "Lucretius is not thinking about an observed uniformity in nature, but rather of the limits which nature imposes on the growth, life, power, etc., of things."

6. Strauss, "Notes on Lucretius," 136 n.16, citing 1.774, 821 and 2.702–3. Trees are Lucretius's example. Trees grow, of course (cf. 1.808), making Lucretius's belief stranger, but perhaps they grow by accretion of atoms, in the way mineral formations may be imagined to grow.

7. Ibid., 121 on 3.295, 5.1325.

8. Democritus's fragment B 278, with Leo Strauss, "The Liberalism of Classical Political Philosophy," in *Liberalism Ancient and Modern* (New York: Basic Books, 1968), 51. On the important question of human free will, Lucretius was closer to Democritus than Epicurus, while still perhaps ranking the latter philosopher above the former; see 3.371, 1039–1041, with Strauss, "Notes on Lucretius," 96, 108, 136 n.8.

9. Alexis de Tocqueville, *Democracy in America*, trans. Harvey C. Mansfield and Delba Winthrop (Chicago: University of Chicago Press, 2000), 676.

10. 4.1026–1287. John Colman, *Lucretius as Theorist of Political Life* (New York: Palgrave MacMillan, 2012), 91. Colman raises the question whether romantic love might not be naturally connected to sex: eros is natural, just as fear of death is. If so, then Lucretius's advice to live without both eros and fear could be seen as a rebellion against nature, or against one of the two natures, the "seen" nature.

11. See, e.g., Francis Bacon, Essay 13, "Of Goodness and Goodness of Nature." Strauss speculates that Lucretius propagates Epicureanism in Rome to secure the protection of powerful political men ("Notes on Lucretius," 107). Martin Ferguson Smith ("Introduction," in *Lucretius: On the Nature of Things* [Indianapolis, IN: Hackett, 2001], xvii) presents evidence that this project backfired in the case of the poem's addressee, Memmius.

12. 1.150 with Epicurus, *Letter to Herodotus*, 38.

13. See also 6.703–711. "His book is entitled not 'On Nature' (as so many Greek books were) but *De rerum natura*, which we should perhaps translate 'On the Naturalness of Things'" (Klein, "On the Nature of Nature," 222).

14. Contrast James Nichols, "On Leo Strauss' 'Notes on Lucretius,'" in *Brill's Companion to Leo Strauss' Writings on Classical Political Thought*, ed. Timothy Burns (Leiden: Brill, 2015), 77, 93. Colman, *Lucretius as Theorist*, 102: "As Lucretius makes clear, the acceptance of materialist principles is a necessary preliminary to acceptance of the mortality of the soul (III, 31–38)."

15. Smith, "Introduction," xxiii.

16. Cf. Strauss, "Notes on Lucretius," 86–87; contrast 96.

17. Ibid., 130–31: Lucretius's "official" teaching is that the most lovable things of all, the gods, are eternal, a teaching which sweetens the sad truth that the world is not of divine origin. Martin Ferguson Smith (Lucretius, *De Rerum Natura*, xxxvii) makes Epicurus "a firm believer in the existence of gods" and writes that "the existence of the gods is certain, for our knowledge of them is derived from clear perception," citing the *Letter to Menoeceus* 123. Strauss's overall study (see especially 114–115) utterly demolishes, on internal evidence, the possibility that Lucretius, at least, believes any gods to be "known" (130; 138 n.87).

18. Strauss, "Notes on Lucretius," 77, 86, 122, 131.

19. *Summa Theologica* I, Q.20 a.1, Q.19 a.2.

20. Hans Jonas, "Biological Foundations of Individuality," *International Philosophical Quarterly* 8 (June 1968): 231–251. Cf. Plato, *Symposium* 207c–208b.

21. *Letter to Menoeceus*, 132.

22. *Principal Doctrines* 11, quoted from Richard Ferrier, "Lucretius' Philosophical Poetry," Annapolis, MD, St. John's College Library Faculty File, 3 (different emphases added).

23. *Symposium* 208c–209e, as contrasted with 211c–212a, particularly the conditional closing sentence.

24. Paul Rahe, "In the Shadow of Lucretius: The Epicurean Foundations of Machiavelli's Thought," *History of Political Thought* 28, no. 1 (2007), 51–53, on Niccolò Machiavelli, *Discourses on Livy*, trans. Harvey C. Mansfield and Nathan Tarcov (Chicago: University of Chicago Press, 1996), 213–14, 21–23.

25. 3.995ff.; cf. James Nichols, *Epicurean Political Philosophy* (Ithaca, NY: Cornell University Press, 1976), 207–210. For Rousseau's renditions of the same idea, see especially the penultimate section of Arthur Melzer's contribution in this volume.

Chapter 10. Jean-Jacques Rousseau

1. Parenthetical citations of the works of Rousseau refer to the following editions: *Lettres Écrites de la Montagne*, in *Oeuvres Complètes*, ed. Bernard Gagnebin and Marcel Raymond, vol. 3 (Paris: Gallimard, Bibliothèque de la Pléiade, 1959–1969); *Letter to M. d'Alembert on the Theatre*, trans. Allan Bloom, in *Politics and the Arts* (Ithaca, NY: Cornell University Press, 1968); *On the Social Contract*, in *On the Social Contract with Geneva Manuscript and Political Economy*, ed. Roger D. Masters, trans. Judith R. Masters (New York: St. Martin's Press, 1978); *Emile: Or on Education*, trans. Allan Bloom (New York: Basic Books, 1979); "Preface to Narcissus: Or the Lover of Himself," in *Discourse on the Sciences and Arts, and Polemics*, vol. 2 of *The Collected Writings of Rousseau*, ed. Roger D. Masters and Christopher Kelly (Hanover, NH: University Press of New England, 1992), 186–198; "Preface to a Second Letter to Bordes," in *Discourse on the Sciences and Arts, and Polemics*, 182–185; "Final Reply," in *Discourse on the Sciences and Arts, and Polemics*, 110–129.

2. It is not possible to try to substantiate this claim here. For a brief effort, see Arthur M. Melzer, *The Natural Goodness of Man: On the System of Rousseau's Thought* (Chicago: University of Chicago Press, 1990), 30 n.1; and Melzer, "The Origin of the Counter-Enlightenment: Rousseau and the New Religion of Sincerity," *American Political Science Review* 90, no. 2 (June 1996): 355.

3. Like the Rousseauian rejection of the conquest of nature, the Romantic opposition to it also did not stem from a sincerely held religion of nature. Still, the true grounds of Rousseau's views, I will be arguing, were essentially classical and so, in most respects, quite different from those of the Romantics. Regarding Romanticism on these and related issues, see Paul Cantor, *Creature and Creator: Myth-Making and English Romanticism* (Cambridge: Cambridge University Press, 1984); and Paul Cantor, "Romanticism and Technology: Satanic Verses and Satanic Mills," in *Technology in the Western Political Tradition*, ed. Arthur M. Melzer, Jerry Weinberger, and M. Richard Zinman (Ithaca, NY: Cornell University Press, 1993), 109–128.

4. For an excellent discussion of Rousseau's scientific work, see Christopher Kelly, "Rousseau's Chemical Apprenticeship," in *Rousseau and the Dilemmas of Modernity*, ed. Mark Hulliung (New Brunswick: Transaction Publishers, 2015), 3–29.

5. Thomas Hobbes, *Leviathan: Or the Matter, Forme, and Power of a Commonwealth Ecclesiasticall and Civil*, ed. Michael Oakeshott (London: Collier-Macmillan, 1962).

Chapter 11. Kant on Organism and History

1. Broadly speaking, the version of mastery Kant endorses is Baconian-Cartesian, and his criticisms of mastery are Rousseauian. This formulation, however, should be seen as only preliminary.

2. L. Strauss, *Thoughts on Machiavelli* (Glencoe, IL: Free Press, 1958), 296.

3. L. Strauss, *What is Political Philosophy? And Other Studies* (New York: The Free Press, 1959), 55. One wonders how one can square this assertion about "oblivion" with the concerns with eternity in Spinoza, Hegel, and Nietzsche, or with Kant's postulate of immortality. Heidegger, for that matter, reflects in later writings on the need to move beyond the horizon of time in order to think the ground of time. See *Zur Sache des Denkens* (Tübingen: M. Niemeyer, 1969). On Spinoza, Strauss himself writes that he "attempts to restore the traditional conception of contemplation: one cannot think of conquering nature if nature is the same as God. Yet Spinoza restored the dignity of speculation on the basis of modern philosophy or science, of a new understanding of 'nature.' He was thus the first great thinker who attempted a synthesis of pre-modern (classical-medieval) and of modern philosophy." *Spinoza's Critique of Religion*, trans. E. Sinclair (New York: Schocken, 1965), 15–16.

4. L. Strauss, *Natural Right and History* (Chicago: University of Chicago Press, 1953), 176.

5. "Of our own person we will say nothing. But as to the subject matter with which we are concerned, we ask that men think of it not as an opinion but as a work; and consider it erected not for any sect of ours, or for our good pleasure, but as the foundation of human utility and greatness . . . Further, each may well hope from our instauration that it claims nothing infinite, and nothing beyond what is mortal; for in truth it prescribes only the end of infinite errors, and this is a legitimate end." Kant, *Critique of Pure Reason* (henceforth *CPuR*), Bii. I have slightly modified the translation by P. Guyer and A. Wood (Cambridge: Cambridge University Press, 1998).

6. *Kants gesammelte Schriften*, Akademie Ausgabe (Berlin: Walter de Gruyter, 1902–), henceforth cited as *KgS*, 20:44 (*Remarks in the* Observations on the Feeling of the Beautiful and Sublime, 1764–65).

7. "Idea for a Universal History with a Cosmopolitan Aim" (henceforth IUH), ninth thesis; *KgS* 8:30.

8. *CPuR* A839–840/B867–868.

9. Ibid., Bxxx. I employ the translation of the *Critique* by Norman Kemp-Smith (New York: St. Martin's Press, 1965). The criticism of reason's theoretical powers restricts knowledge of objects to the realm of spatio-temporal appearances and allows, beyond the realm of appearances, the postulations of ideas of God, freedom, and immortality as supports of practical reason.

10. *CPuR* Bxv.

11. Ibid., A797/B825. See B395, note: "Metaphysics has as the proper objects of its inquiries three ideas only: God, freedom and immortality."

12. Ibid., Bxxiii.

13. Ibid., B23.

14. Ibid., Axx; also B26.

15. I need not say much about how this reading of Kant differs from readings based on the thought of John Rawls, which are still very influential in American academic philosophy.

16. *Religion within the Boundaries of Mere Reason* (henceforth *R*), Part One; "The End of All Things," *KgS* 8:333–335.

17. IUH, sixth thesis.

18. *R* 6:59: "For if all the world proceeded in accordance with the precept of the law, we would say that everything occurred according to the order of nature, and nobody would think even of inquiring after the cause."

19. *CPuR* A852–856/B880–884, "The History of Pure Reason": "The *critical* path alone is still open . . . .[The reader] may now judge for himself whether, if he cares to lend his aid in making this path into a high-road, it may not be possible to achieve before the end of the present century what many centuries have not been able to accomplish."

20. *CPuR* B21: "Thus in all men, as soon as their reason is ripe for speculation, there has always existed and will *always continue to exist* some kind of metaphysics" (my emphasis); "Plato very well realized that our faculty of knowledge feels a much higher need than merely to spell out appearances according to a synthetic unity. . . . He knew that our reason naturally exalts itself to modes of knowledge which so far transcend the bounds of experience" (ibid. A314/B370–371); "The absolute whole of appearances—we might thus say—*is only an idea*; since we can never represent it in image, it remains a *problem* to which there is no solution" (ibid. A328/B384).

21. *CPuR* Bxii. Kant here ascribes a modern conception of mathematicals to the ancient discovery. For the contrast between ancient "ontological" and modern symbolic-constructive accounts of mathematics, see J. Klein, *Greek Mathematical Thought and the Origin of Algebra*, trans. E. Brann (Cambridge, MA: MIT Press, 1968) and D. Lachterman, *The Ethics of Geometry* (New York: Routledge, 1989).

22. Ibid., Bxiii.

23. I refer to Richard Kennington's superb account of the new approach to nature in terms of laws rather than forms, species, and beings. See my summary of the account in "Masks of Mastery: Richard Kennington on Modern Origins," *Political Science Review* 31 (2002), 6–28. This should not replace Kennington's own studies in *On Modern Origins: Essays in Early Modern Philosophy* (Lanham, MD: Lexington Books, 2004).

24. See *Rules* 3 and the account in *Discourse on Method*, part 2. Also see the discussion of Pamela Kraus, "Introduction," in R. Descartes, *Discourse on Method*, trans. R. Kennington (Newburyport, MA: Focus, 2007), 1–13.

25. *Rules* 2 and 13.

26. *Discourse on Method*, part 2. A comparable move can be found in Spinoza, *Ethics*, part 4, where the account of universal conatus in bodies is supplemented and modified by the construction of models of human nature.

27. See *Meditations on First Philosophy*, VI.

28. The most common approach to Descartes reads him as founding "epistemology" through seeking to overcome skepticism.

29. This is the amazing failure of Bernard Williams's exposition of Descartes's "project of pure inquiry."

30. For an excellent discussion of Descartes's ultimate project as the development of a therapeutic "highest and most perfect moral science" of the passions, see Richard Hassing, *Cartesian Psychophysics and the Whole Nature of Man* (Lanham, MD: Lexington Books, 2015).

31. See M. Friedman, *Kant and the Exact Sciences* (Cambridge, MA: Harvard University Press, 1992) for Kant's revision of Newton's conceptions of space, time, and force.

32. *CPuR* A724–725/B752–753.

33. Kant cites a line from Ovid depicting original chaos, *Metamorphoses*, I.16: *instabilis tellus, innabilis unda.*

34. *CPuR* A752/B780.

35. See Susan Shell's discussion of modern organism as uniting ancient meanings of organon (Latin: *organum*) as bodily part and tool or instrument, in a living system: "An organism is a whole made up of interdependent, tool-like parts that function as tools or instruments of one another." S. M. Shell, "Organizing the State: Transformations of the Body Politic in Rousseau, Kant, and Fichte," in *Der Begriff des Staates. Internationales Jahrbuch des Deutschen Idealismus*, ed. K. Ameriks and J. Stolzenberg (Berlin: W. de Gruyter, 2004), 53. On Kant's account of reason as organism see A. Ferrarin, *The Powers of Pure Reason: Kant and the Idea of Cosmic Philosophy* (Chicago: University of Chicago Press, 2015).

36. *Critique of the Power of Judgment* (henceforth *CPJ*), Introduction, sections II and IX. I use the translation of P. Guyer and E. Matthews (Cambridge: Cambridge University Press, 2000). Although the power of judgment (*Urteilskraft*) lacks a legislative domain, it has a realm of application (*Anwendung*) in art (*Kunst*), *KgS* 5:198. This points to the crucial fact that art and purpose are common to both parts of the *Critique.*

37. H. Allison, "Teleology and History in Kant: the Critical Foundations of Kant's Philosophy of History," in *Kant's 'Idea for a Universal History with a Cosmopolitan Aim': A Critical Guide*, ed. A. O. Rorty and J. Schmidt (Cambridge: Cambridge University Press, 2009). Allison discusses how the treatment of teleology in the third *Critique* relates to earlier treatments of judgment and purpose in Kant's writings.

38. J. Zammito, *The Genesis of Kant's Critique of Judgment* (Chicago: University of Chicago Press, 1992).

39. *KgS* 5:174–179.

40. Ibid., 5:179.

41. Ibid., 5:360. Kant discusses the need to think in terms of purposes by analogy with art as "a peculiarity of the human understanding" in *CPJ*, sections 76–77. For a discussion, see the author's *Being After Rousseau: Philosophy and Culture in Question* (Chicago: University of Chicago Press, 2002), 93–109.

42. *KgS* 5:362–366. Kant sets aside the case of the purposiveness for our understanding of mathematical objects, which is only a purposiveness that we introduce into figures we construct; it is merely "formal," as it does not involve beings materially independent of the human mind.

43. Ibid., 5:367–369.

44. Ibid., 5:370–372; also 384.

45. Ibid., 5:372–374.

46. Ibid., 5:374–376.

47. Aristotle, *De Anima.* Pertinent here is the observation of Rachel Zuckert: "Kant places emphasis not on the purpose of the object, i.e., the good it serves, the reason why it exists or why a rational agent created it, but, rather, on the *kind* of order a purpose constitutes among parts or properties of an object, i.e., an order of diversity and of contingency." R. Zuckert, *Kant on Beauty and Biology: An Interpretation of the Critique of Judgment* (Cambridge: Cambridge University Press, 2007), 10.

48. It is worth comparing this to the emphasis on botanical study in Rousseau and Goethe.

49. *KgS* 5:380.

50. Ibid., 5:376.

51. Ibid., 5:400, also 410.

52. Ibid., 5:411, 415.

53. Ibid., 5:376.

54. Ibid., 5:380. "And we can love [nature] for this, just as we regard it with respect because of its immensity, and can feel ourselves ennobled in this contemplation—just as if nature had erected and decorated its magnificent stage precisely with this intention."

55. Ibid., 5:413–415.

56. Ibid., 8:18 (IUH, first thesis). Of human natural capacities, it must be said, "They are destined to develop completely and in conformity with their end. . . . In the teleological theory of nature, an organ that is not intended to be used, an organization that does not achieve its end, is a contradiction."

57. Ibid., 5:380.

58. Ibid., 5:426–427.

59. Ibid., 5:431.

60. Ibid., 5:435.

61. Ibid., 5:429–430. Nature's ultimate purpose is the theme of *CPJ*, section 83.

62. Ibid., 5:431.

63. Ibid., 5:436, note.

64. Ibid., 5:430.

65. Ibid., 5:433. See the related account of "unsocial sociability" and its role in the development of reason (the arts, sciences, and the formation of political institutions) in IUH. See also *Toward Perpetual Peace.*

66. *KgS* 5:433–434. See also *CPJ*, section 60.

67. Kant draws on Rousseau's analysis of the corruptions of culture (the arts and sciences), but he has, unlike Rousseau, confidence or hope in the self-correcting trajectory of their development.

68. See *CPuR*, "The Architectonic of Pure Reason," A832–851/B860–879, and *Anthropology from a Pragmatic Standpoint*, section 57, *KgS* 7:224–227.

69. Also the core issue in Fichte, Schelling, and arguably (contrary to his protests) Hegel.

Chapter 12. Beyond the Island of Truth

1. Hegel, *Werke in zwanzig Bänden* [henceforth *Werke*], ed. Eva Moldenhauer and Karl Markus Michel, 20 vols. (Frankfurt a. M.: Suhrkamp, 1970–1971), 20:359; on the reactions of Kant's immediate successors, see Frederick C. Beiser, *The Fate of Reason: German Philosophy from Kant to Fichte* (Cambridge, MA: Harvard University Press, 1987).

2. See Robert Pippin, *Hegel's Idealism: The Satisfactions of Self-Consciousness* (Cambridge: Cambridge University Press, 1989).

3. *Werke* 6:573. The object itself, *Sein*, "being," or nature, is at the same time recognized as the repository of reason as it exists in itself but not for itself.

4. Ibid., 5:61; 9:20.

5. Ibid., 5:61.

6. On this point, see Amos Funkenstein, *Theology and the Scientific Imagination from the Middle Ages to the Seventeenth Century* (Princeton, NJ: Princeton University Press, 1986).

7. "Introduction," in *Hegel and the Philosophy of Nature*, ed. Stephen Houlgate (Albany: State University of New York Press, 1998), xi.

8. A list of the critics of Hegel's natural philosophy can be found in Sebastian Rand's "The Importance and the Relevance of Hegel's 'Philosophy of Nature,'" *Review of Metaphysics* 61, no. 2 (December 2007): 381.

9. If we think reductionism to its conclusion, there is only one real science, physics, and the Aristotelian notion that each science has its own first principles becomes untenable.

10. Or as Hegel puts it, if nature is alienated from the idea, it is merely the corpse of the understanding. *Werke* 9:25. Nature for Hegel is a living whole.

11. Brigitte Falkenburg, "How to Save the Phenomena: Meaning and Reference in Hegel's Philosophy of Nature," in *Hegel and the Philosophy of Nature*, 109.

12. *Werke* 7:24.

13. Falkenburg, 102.

14. Ibid., 108.

15. Ibid., 98; see also *Werke* 5:29.

16. "Introduction," xix.

17. Falkenburg, 98.

18. Ibid., 113, 126.

19. Dieter Wendschneider, "Die Absolutheit des Logischen und des Sein der Natur. Systematische Überlegungen zum absolut-idealistischen Ansatz Hegels," *Zeitschrift für philosophische Forschung* 39, no. 3 (July–September, 1986): 349.

20. William Maker, "The Very Idea of Nature or Why Hegel Is Not an Idealist," in *Hegel and the Philosophy of Nature*, 45.

21. "The Importance and the Relevance of Hegel's 'Philosophy of Nature,'" 379–400. Rand describes how he imagined this could be achieved. In what follows I draw on his argument.

22. Rand, 391.

23. Hegel here distinguishes between will (*Wille*) and caprice (*Wilkür*). Will is a rationalized form of caprice that does not just want, but wants what is good and rational.

24. *Werke* 9:16.

25. Ibid., 6:549.

26. Ibid., 10:18–19.

27. Michael A. Gillespie, *Hegel, Heidegger, and the Ground of History* (Chicago: University of Chicago Press, 1984), 100–103.

28. What is *simply* wrong in Hegel's account—for example, his rejection of evolution or his assertion that there can be no planet between Mars and Jupiter—is accidentally wrong, not the result of the needs of his system but of the defects of the science of his time. It is thus not inconceivable that he could have constructed his natural philosophy in a different manner if he had been aware of what we now know. These errors, then, do not necessarily undermine his system of science.

29. *Werke* 6:573.

30. J. N. Findlay, *Hegel: A Re-examination* (New York: Macmillan, 1958), 269.

31. *Werke* 5:44.

32. Ibid., 9:23.

33. *Kritik der praktischen Vernunft*, in *Gesammelte Schriften* (Berlin: der Königlich-Preussischen Akademie der Wissenschaften zu Berlin, 1902–), I 5:162.

34. Friedrich Nietzsche, *Werke: Kritische Gesamtausgabe*, ed. G. Colli and M. Montinari, (Berlin: de Gruyter, 1967), V 1:335.

Chapter 13. Separating the Moral and Theological Prejudices and Taking Hold of Human Evolution

1. Friedrich Nietzsche, *On the Genealogy of Morals and Ecce Homo*, trans. Walter Kaufmann (New York: Vintage Books, 1989), 17. All references to this book (*GM*) will be included in the body of the text with the appropriate section number.

2. Nietzsche, "Beyond Good and Evil," in *Basic Writings of Nietzsche*, trans. and ed. Walter Kaufmann (New York: The Modern Library, 1992), 192–194. All references to this book (*BGE*) will be included in the body of the text with the appropriate section number.

3. Nietzsche, *The Gay Science*, trans. Walter Kaufmann (New York: Vintage Books, 1974), 79. All references to this book (*GS*) will be included in the body of the text with the appropriate section number.

4. What is true of differences between species is true at the more precise level of the individual. We have many experiences that are similar enough to allow us to conclude we are closely related to other human beings, but no individual is physio-psychologically exactly the same as any other. We might try to enter more fully into the experience of another individual, but, so long as our physio-psychologies are not identical, our experiences will not be the same. This line of reasoning can be extended to the individual. Experiences an individual has at one time can also never be repeated exactly, unless this individual's physio-psychology is exactly the same at both times.

5. Nietzsche, *Daybreak*, trans. R. J. Hollingdale, ed. Maudemarie Clark and Brian Leiter (Cambridge: University of Cambridge Press, 1997), 80–82; and *Twilight of the Idols and the Anti-Christ*, trans. R. J. Hollingdale (London: Penguin Books, 1990), 50–51. All references to these books, (*D*) and (*TI*), will be included in the body of the text with the appropriate section numbers. This is not to say that only things we can know or that are available to us can have consequences for us. It is to say we cannot know anything of such causes; hence, we cannot recognize their effects for what they are. Speaking in moral terms, this means there might be an unknowable being who punishes us for certain actions, but, since we have no access to such a being and cannot recognize the punishments for what they are, we cannot mindfully adjust our behavior so as to avoid punishment. At best, we could modify our behavior because we see patterns in our actions and their effects, but this adjustment would not be morally motivated. We would not change our behavior because we know it to be evil. Since we can know nothing of this kind of supra-human being or the possible ills it might impose on us, they fall outside the realm of things that should properly attract our attention and interest.

6. The overthrow of these theological and moral prejudices does not mean all concepts of either the divine or of morality are disproved. The overthrow is specific to the prejudices as we have defined them.

7. Nietzsche makes an analogous observation when reflecting on the difference between his thinking and what he articulates in words. Thinking is fluid, but this fluidity necessarily ceases as soon as one tries to capture the thoughts in words. Nietzsche maintains his thinking is more beautiful, more enlivening, than his written words. He laments this difference between his thinking and his written articulations of it (*BGE* 296).

8. Nietzsche, *Human, All Too Human*, trans. R. J. Hollingdale (Cambridge: Cambridge University Press, 1996), 35. All references to this book (*HH*) will be included in the body of the text with the appropriate section number.

Chapter 14. Mastery of Nature and Its Limits

Numbers in parentheses for Heidegger, *Introduction to Metaphysics*, refer to the German, and are included in the English edition to which I refer.

1. Consider the several divisions of Plato's *Statesman*.

2. Consider Jacob Klein, *Greek Mathematics and the Origin of Algebra*, trans. Eva Brann (Cambridge, MA: MIT Press, 1968); Martin Heidegger, *What is a Thing*, trans. W. B. Barton Jr. and Vera Deutsch (Chicago: Henry Regnery, 1967), originally published as *Die Frage Nach Dem Ding* (Tubingen: Niemeyer, 1962); and Martin Heidegger, *The Question Concerning Technology*, trans. William Lovitt (New York: Harper & Row, 1977), originally published as *Die Frage Nach dem Technik*, in *Vortrage und Aufsatze* (Pfullingen: Neske, 1954).

3. "Allein, wo ist entschieden, dass die Natur als solche fur alle Zukunft die Natur der modernen Physik bleiben . . . ?" Martin Heidegger, *Identity and Difference*, trans. Joan Stambaugh (New York: Harper & Row, 1969), 105, originally published as *Identitat und Differenz* (Pfullingen: Neske, 1957).

4. Consider Martin Heidegger, *Nietzsche*, ed. David Farrell Krell, trans. Frank Capuzzi, David Farrell Krell, and Joan Stambaugh (New York: Harper & Row, 1979–1987), originally published in German by Pfullingen: Neske (1961).

5. Consider Martin Heidegger, *Being and Time*, trans. John Macquarrie and Edward Robinson (New York: Harper & Row, 1962). Macquarrie and Robinson's translation is of the seventh edition of *Sein und Zeit* (Tubingen: Niemeyer, 1953). *Sein und Zeit* was originally published in 1927.

6. Consider *Identity and Difference*: "Wir durfen aber noch weniger der Meinung nachhangen, die technische Welt sei von einer Art, die einen Absprung aus ihr schlechthin verwehere" (105). Nonetheless, "Zwar konnen wir die heutige technische Welt weder als Teufelswerk verwerfen."

7. Consider Martin Heidegger, *Introduction to Metaphysics*, trans. Gregory Fried and Richard Polt, 2nd ed. (New Haven, CT: Yale University Press, 2014), 41 (28), originally published as *Einfuhrung in die Metaphysik*, (Tubingen: Niemeyer, 1953).

8. See *Introduction to Metaphysics*, 222 (152).

9. If, for example, we strive in the atomic age only for the peaceful use of atomic energy, we secure the metaphysical mastery of the technological world. "So lange die Besinnung auf die Welt des Atomzeitalters bei allem Ernst der Verantwortung nur dahin drangt, aber auch nur dabei als dem Ziel sich beruhigt, die friedliche Nutzung der Atomenergie zu betreiben, so lange bleibt das Denken auf halbem Wege stehen. Durch diese Halbheit wird die technische Welt in ihrer metphysischen Vorherrschaft weiterhin und erst recht gesichert." *Identity and Difference*, 105.

10. See *Being and Time*, division I, chapter VI.

11. From Plato through Hegel our imperfection is recognized and with Hegel presumably overcome, only to return in its way with Nietzsche and Heidegger.

12. See Mark Blitz, *Heidegger's* Being and Time *and the Possibility of Political Philosophy*, with a new afterword (Philadelphia: Paul Dry Books, 2017).

Chapter 15. What Is Natural Philosophy?

1. I am neglecting here certain indications in the writings of Xenophon that Socrates did not in fact limit his inquiries to the human things or abandon the study of nature as a whole—as, for example, when Xenophon writes, in *Memorabilia* 4.6.1, that Socrates "never ceased considering what each of the beings is."

2. David Bolotin, *An Approach to Aristotle's Physics* (Albany: State University of New York Press, 1998).

3. Whether Bolotin's way reading of Aristotle's *Physics* is correct is an open question; in my opinion, Bolotin is not wrong about the rhetorical character of Aristotle's writing in the *Physics*, but in deciding which of Aristotle's surface teachings are merely exoteric and not meant as serious natural philosophy, he sometimes throws the baby out with the bathwater.

4. Albert Einstein and Leopold Infeld, *The Evolution of Physics* (Cambridge: Cambridge University Press, 1938).

5. Albert Einstein, "Zur Elektrodynamik bewegter Körper," *Annalen der Physik* 17:891 (1905); published in English as "On the Electrodynamics of Moving Bodies," in *The Principle of Relativity*, trans. W. Perrett and G. B. Jeffery (London: Methuen, 1923), 35–65.

6. Hermann Minkowski, "Raum und Zeit," a lecture at the 80th Meeting of German Scientists and Physicians, Cologne, September 21, 1908; published in English as "Space and Time," in Perrett and Jeffery, *The Principle of Relativity*, 73–91.

7. Minkowski, "Space and Time," 83.

8. See Abraham Pais, *Subtle is the Lord: The Science and Life of Albert Einstein* (Oxford: Oxford University Press, 1982), chapter 9.

9. See Richard Feynman, *QED: The Strange Theory of Light and Matter* (Princeton, NJ: Princeton University Press, 1985).

Chapter 16. Quantum Mechanics and Political Philosophy

I would like to thank Wayne Ambler, Ewa Atanassow, Mark Blitz, Daniel Doneson, Alberto Ghibellini, Peter Hansen, Hugh Liebert, Steven Lenzner, Paul Ludwig, Svetozar Minkov, Rory Schacter, Adam Schulman, Nathan Tarcov, and Peter Thiel for most helpful discussions, critiques, suggestions, and insights.

1. M. Planck, "Zur Theorie des Gesetzes der Energieverteilung im Normalspectrum," *Verhandlungen der Deutschen Physikalischen Gesellschaft* 2 (1900): 237; English translation "On the Theory of the Energy Distribution Law of the Normal Spectrum," from *The Old Quantum Theory*, ed. D. ter Haar (Oxford: Pergamon Press, 1967), 82.

2. P. A. M. Dirac, *Proceedings of the Royal Society A* 123 (1929): 714.

3. It is important to note that quantum mechanics does not lead to the introduction of new forces, but only a new way of treating forces.

4. Here I focus on the significance of quantum mechanics per se, while using a few of the insights of quantum field theory, such as the fundamental particles.

5. One is the charming Max Tegmark, *Our Mathematical Universe* (New York: Vintage Books, 2014), with a clarity of presentation in non-mathematical terms and an aim towards comprehensiveness. There is also a large bibliography in it, and a search would reveal many other studies; cf. T. L. Pangle, "On Heisenberg's Key Statement Concerning Ontology," *Review of Metaphyslcs* 67 (2014): 835.

6. In quantum mechanics, there is intrinsic, non-reducible randomness in each event. A large number of the same type of event follows a strict pattern, but for a given event, e.g., the directions the particles go after they interact, there is a randomness which simply is. Also, particles interact non-locally through space in a way different from action at a distance. This nonlocality, similarly, simply is.

7. Aristotle, *Physics*, 184a1–2.

8. This is not to say that eventually these elements might not be found to themselves consist of smaller elements, but that within the framework of quantum mechanics as such, they are indivisible.

9. I write "in principle" because, of course, not everything has been explained. There is a tremendous amount of work yet to do, but that work involves the application of quantum mechanical principles, not the development of new principles. Of course, modern physics cannot state with certainty that new principles governing the sub-astronomic will not be found, but with its billions of successful applications in explaining the hitherto unexplained, there is no reason to assume that new principles are needed. Having said that, it should also be emphasized that the "fundamental" forces and "universal" constants, not to mention the patterns in nature, cannot be explained as such. They must be taken as given, irreducible aspects of nature that simply are.

10. I am well aware of the huge body of literature on quantum mechanics and consciousness and the many unsettled debates on this subject. The self-imposed criterion of the ultimate success of modern physics is that it can explain everything based on the principles elucidated by the application of its method. To the extent that it cannot do that (for example, because of irreducible heterogeneity), it is not successful.

11. See, for example, F. Wilczek, *A Beautiful Question* (New York: Penguin Press, 2015), but also in this book is an openness to the heterogeneity of things, as discussed below.

12. As noted above, I use "quantum mechanics" here as incorporating key insights from quantum field theory. I also focus on the microscopic. General relativity would lead to truth about the astronomic, and TOE, about all scales.

13. Plato, *Theatetus*, 173c7–d2; I refer to the translation by S. Benardete (Chicago: University of Chicago Press, 1986).

14. I just made a very rough estimate by assuming a 70 kg. person (a standard assumption in pharmaceuticals) composed of water. Using a 6-31g(2d,2p) with a DFT functional, an optimization of a water molecule took about 1 minute on my laptop. I assumed a 10^6 more powerful computer and N^2 scaling. This number is 10 undecillion.

15. This is actually the solution to the problem of quantum mechanical measurements called superdeterminism. On the other side of the coin, quantum mechanics itself is just a product of random processes in the brains of its developers, which are themselves just a product of random processes. Either way, mechanism implodes on itself: it is an illusion or of no real significance. Nature does not let herself be so easily mastered, even theoretically.

16. Irrational numbers might seem like an exception (or the norm given that there is an infinite number of them for each rational number), since enumerated, they would create an infinity of randomness. However, there is no way to actually enumerate them within the realm of mathematics.

17. See P. Galison, "Kuhn and the Quantum Controversy," *British Journal for the Philosophy of Science* 32 (1981): 73.

18. M. Planck, *Eight Lectures on Theoretical Physics* (New York: Columbia University Press, 1915), 6–7. Planck first delivered these lectures in 1909.

19. Lucretius, *On the Nature of Things*, I.157–158.

20. See Ludwig Boltzmann's lectures in *Theoretical Physics and Philosophical Problems*, ed. Brian McGuinness (Boston: D. Reidel, 1974), 13, 32, 75, and the last paragraph of 198. See also Hans Reichenbach, *Philosophic Foundations of Quantum Mechanics* (Berkeley: University of

California Press, 1944), vii. Reichenbach states, "The author has tried in the present book to develop a philosophical interpretation of quantum physics which is free from metaphysics." It is unclear whether Reichenbach is equating metaphysics with divinity, lack of absolutism, or both.

21. Plato, *Theaetetus*, 201D 1–4.

22. See page 612 in P. A. M. Dirac, "The Quantum Theory of the Electron," *Proceedings of the Royal Society of London. Series A*, 117 (1928): 610. 612. The meaning of the symbols can be found in the reference.

23. Tegmark, *Our Mathematical Universe*, and Wilczek, *A Beautiful Question*, respectively.

24. Wilczek, *A Beautiful Question*, 29.

25. L. Strauss, *What Is Political Philosophy? And Other Studies* (Chicago: University of Chicago Press, 1959), 11.

26. This example itself, though, implies that the otherwise non-political is more political than the otherwise political is non-political; for justice, even if considered a product of mechanism, still retains its political significance, even if that significance is secondary to mechanism, whereas an atom could be viewed fully mechanistically, with no political significance.

List of Contributors

ROBERT C. BARTLETT is the Behrakis Professor of Hellenic Political Studies at Boston College.

MARK BLITZ is the Fletcher Jones Professor of Political Philosophy at Claremont McKenna College.

DANIEL A. DONESON is a Lecturer in the Department of Chemical Engineering at the Massachusetts Institute of Technology.

MICHAEL A. GILLESPIE is the Jerry G. and Patricia Crawford Hubbard Professor of Political Science at Duke University.

RALPH LERNER is the Benjamin Franklin Professor Emeritus at The College and the Committee on Social Thought, University of Chicago.

PAUL LUDWIG is a tutor at St. John's College, Annapolis.

HARVEY C. MANSFIELD is Professor of Government at Harvard University.

ARTHUR MELZER is Professor of Political Science at Michigan State University.

SVETOZAR Y. MINKOV is Associate Professor of Philosophy at Roosevelt University.

CHRISTOPHER NADON is Associate Professor of Government at Claremont McKenna College.

DIANA J. SCHAUB is Professor of Political Science at Loyola University Maryland.

ADAM SCHULMAN is a tutor at St. John's College, Annapolis.

DEVIN STAUFFER is Associate Professor of Government at the University of Texas at Austin.

BERNHARDT L. TROUT is the Raymond F. Baddour, ScD, (1949) Professor of Chemical Engineering at the Massachusetts Institute of Technology.

LISE VAN BOXEL is Professor of Philosophy at St. John's College, Santa Fe and Annapolis.

RICHARD VELKLEY is the Celia Scott Weatherhead Professor of Philosophy at Tulane University.

STUART D. WARNER is Associate Professor of Philosophy at Roosevelt University and Director of the Montesquieu Forum.

JERRY WEINBERGER is University Distinguished Professor Emeritus at Michigan State University.

Index

Acknowledgments

As editors, we naturally have many people to whom we owe our gratitude. Above all, we are grateful to all the writers who were unfailingly professional and whose contributions are an intellectual delight. Since the book grew out of the Mastery of Nature Conference at MIT, May 12–13, 2016, we would also like to thank the chairs and commentators at that event: Bryan Garsten, Ralph Lerner, Susan Shell, Steven Smith, Vickie Sullivan, and Nathan Tarcov. Heartfelt thanks to Daniel Doneson, who was a most helpful partner in the conception and development of the conference. Thanks to Alex Limanowski for his help on the index. We are also thankful to all of the participants who came to or engaged in the meeting; in particular, Peter Thiel, who not only had excellent questions and comments throughout, but gave a thought-provoking keynote address. Dr. Damon Linker at the University of Pennsylvania Press saw early on the value of this book and was most helpful in steering us to completion. We would also like to thank the Jack Miller Center and the John Templeton Foundation for their encouragement and financial support.